Building a Multinational Global Navigation Satellite System

An Initial Look

D1713740

Rosalind Lewis, Michael Kennedy,
Elham Ghashghai, Gordon Bitko

Prepared for the United States Air Force

PROJECT AIR FORCE

The research described in this report was sponsored by the United States Air Force under Contract F49642-01-C-0003. Further information may be obtained from the Strategic Planning Division, Directorate of Plans, Hq USAF.

Library of Congress Cataloging-in-Publication Data

Building a multinational global navigation satellite system : an initial look /
    Rosalind Lewis ... [et al.].
        p. cm.
    "MG-284."
    Includes bibliographical references.
    ISBN 0-8330-3735-8 (pbk. : alk. paper)
    1. Artificial satellites in navigation. 2. Global Positioning System.
    I. Lewis, Rosalind.

TL798.N3B85 2005
623.89'3—dc22

                                                                    2005000550

The RAND Corporation is a nonprofit research organization providing objective analysis and effective solutions that address the challenges facing the public and private sectors around the world. RAND's publications do not necessarily reflect the opinions of its research clients and sponsors.

RAND® is a registered trademark.

Published 2005 by the RAND Corporation
1776 Main Street, P.O. Box 2138, Santa Monica, CA 90407-2138
1200 South Hayes Street, Arlington, VA 22202-5050
201 North Craig Street, Suite 202, Pittsburgh, PA 15213-1516
RAND URL: http://www.rand.org/
To order RAND documents or to obtain additional information, contact
Distribution Services: Telephone: (310) 451-7002;
Fax: (310) 451-6915; Email: order@rand.org

Preface

In the not too distant future, there may be a second global space-based positioning, navigation, and timing (PNT) capability similar to the Global Positioning System (GPS). The European Union plans to begin initial operations of the Galileo PNT system in 2008. What effect this additional and highly capable information utility will have on global economic and security conditions is uncertain. Policy leaders and technical experts have been in discussion since 2000 to find cooperative means of providing users the benefit of both systems. However, some U.S. policymakers are concerned that Galileo will be a threat to U.S. economic and security interests.

In March 2002, when its initial funding was made available, Galileo took one step closer to becoming a reality. In August 2002, in response to a National Security Council request, principals of the Interagency GPS Executive Board (IGEB), a policymaking body established in 1996 by Presidential directive to manage GPS and its U.S. government augmentations, developed recommendations for continued discussions between the United States and the European Union. One month later, the Senior Steering Group–International Space Cooperation (SSG-ISC) commissioned a study on the business case and economic impact to the global user community of two systems, GPS and Galileo. The SSG-ISC is the key forum through which the U.S. Air Force/XO dealt with Galileo issues, and AF/XO asked the RAND Corporation to conduct this study. The study was incorporated into the 2002–2003 RAND Project AIR FORCE research agenda.

Galileo, as envisioned, is very similar to GPS in function and performance, and it has the potential to create new PNT standards in addition to the de facto standards that currently exist in GPS. The focus of this study was the economic impact of a competition that could result from the implementation and operation of Galileo in the presence of GPS. The nature of competition, in this study, was defined by three factors: interoperability and compatibility; strategies employed to foster Galileo adoption; and the schedules for GPS modernization and Galileo development. Our primary measure of the economic impact is *net economic benefits to users of PNT products/services*, which are defined as the difference between the users' (consumers') valuation of the products/services provided and the market prices of those products/services. The implications for the United States are linked to the conditions that warrant a U.S. response to a situation or opportunity created by Galileo.

This report should be of special interest to the members of the IGEB, the GPS Industry Council, and policymakers involved in international negotiation and coordination of PNT systems and information. It was prepared for AF/XO within the Aerospace Force Development Program of RAND Project AIR FORCE.

RAND Project AIR FORCE

RAND Project AIR FORCE (PAF), a division of the RAND Corporation, is the U.S. Air Force's federally funded research and development center for studies and analyses. PAF provides the Air Force with independent analyses of policy alternatives affecting the development, employment, combat readiness, and support of current and future aerospace forces. Research is performed in four programs: Aerospace Force Development; Manpower, Personnel, and Training; Resource Management; and Strategy and Doctrine. The research reported here was prepared under contract F49642-01-C-0003.

Additional information about PAF is available on the RAND Website at http://www.rand.org/paf.

Contents

Figures

Tables

Summary

GPS and Galileo (see pp. 13–20)

The Global Positioning System (GPS) has been the preeminent source for positioning, navigation, and timing (PNT) data in many nonmilitary applications, including various modes of transportation. GPS and its U.S. government augmentations are managed by the Interagency GPS Executive Board (IGEB), which was established by Presidential directive in 1996.[1] The IGEB's functions and responsibilities support the U.S. objective of establishing GPS as the standard PNT source for the national and international community. This objective enables the United States to retain control of a critical information technology and ensures that U.S. organizations can actively participate in the economic growth and technical maturity that result from this technology. No other system has presented a credible competitive threat to this objective, until now.[2]

Galileo, a European space-based PNT system, will be similar to GPS in many ways, such as providing a free service for mass-market applications; but it will be very different in other ways, such as having civilian management and control, as well as a fee-for-service compo-

---

[1] http://www.igeb.gov.

[2] The Russian military-operated PNT system, known as the Global Navigation Satellite System (GLONASS), began operating in 1993. However, it has not been maintained well, and aging satellites have not been replaced. Russian officials have announced a development program to increase the constellation size to 18 by 2008 using longer-life satellites ("GLONASS, GPS and Galileo: A Multi-Expert Interview," 2003).

nent. The significance of these similarities and differences partly depends on the user's perspective. For example, consistent spectrum use across both systems would benefit the civilian user but could complicate U.S. military objectives. The dimensions of GPS's and Galileo's coexistence encompass technical, geopolitical, regulatory, national security, and economic issues.

Of the many uncertainties about a future world in which GPS and Galileo coexist, economic impact is the one that implicitly embodies the concerns of some in the GPS civil community and directly challenges the motivations for Galileo. There are concerns that the competitive environment ushered in by Galileo, with its different technical design and management practices, will create a fragmented or shifted (from GPS to Galileo) user base for PNT information and services. The stated motivations for Galileo are to create jobs, to increase market participation of European firms, and to reduce reliance on the United States—motivations that have caused some to view the Galileo competitive approach as more destructive than constructive.[3] Which competitive environment Galileo will present is not yet clear.

When viewed from a broader perspective, competition is seen as a positive condition, even when it reshapes the landscape (Lancop, 1997). And the landscape in aerospace has seen this sort of change before. Certainly the success of the European Space Agency (ESA) in establishing a European presence in launch activities via Ariane and in commercial aircraft via Airbus is enough to give one pause about what Galileo might mean for GPS. Who will benefit and who will pay as a result of the changes ushered in by Galileo?

To explore the economic ramifications of Galileo, we considered a competitive environment in which competition is defined by three factors: interoperability and/or compatibility, strategies employed to foster Galileo adoption, and the schedules for GPS modernization and Galileo development. What influence might these factors have on

---

[3] *Constructive competition* refers to surpassing the competition by providing a superior product/service. It may lead to continual innovation. *Destructive competition* refers to prohibiting, outmaneuvering, or otherwise decimating the competition to create an advantage for one competitor over the other.

the economic impact of GPS and Galileo coexisting? How should the United States respond in anticipation of Galileo, regardless of whether it succeeds or fails?[4]

Study Boundaries (see pp. 9–11)

The complexity of the GPS and Galileo situation necessitated that we set firm boundaries for our assessment of the three factors. For the interoperability and compatibility assessment, we adopted the parameters currently used by the GPS community and then limited our inquiry to considering the ramifications of these parameters, particularly along economic lines. We do not comment technically on Galileo's design, and we make no comparisons intended to rank the two systems.

For the second factor, strategies employed to foster Galileo adoption, we explored the economic ramifications of mandating the use of Galileo (in certain markets) or restricting (industry) opportunities for participating in Galileo, without commenting directly on the soundness of the business model.

For the third factor, GPS modernization schedules and Galileo development, we considered the incremental capabilities offered by GPS and Galileo. Although we note challenges for both efforts in attaining their schedules, we make no prediction about when the enhanced/new capabilities will actually emerge.

We used the PNT industry as a proxy for the user in our assessment of economic benefits because of the industry's inherent connection to the user base. Literature reviews, discussions with domain experts, and industry surveys informed our observations about the competitive factors in areas related to performance, management, and use of satellite PNT. We talked to representatives of the GPS Joint Program Office (JPO), the Office of the Assistant Secretary of Defense for Command, Control, Communications, and Intelligence

---

[4] This study assumes that Galileo succeeds—in other words, that it achieves the advertised capability as planned.

(OASD C3I), members of the Institute of Navigation (ION), and members of the GPS Industry Council. Additionally, we held exploratory discussions with several manufacturers and service providers to probe the significance of GPS and Galileo coexisting and the consequences of the competitive factors.

Collectively, these sources were used to develop, distribute, and evaluate surveys to better understand the potential economic ramifications. Of the approximately 250 contacts we made with industry, only 19 completed the survey, and even with the direct industry contacts, the sample size is not representative. Therefore, the results are illustrative but cannot serve as the basis for generalizations. The companies that we interacted with (either directly or via survey) are listed in Appendix A. Both forms of respondents, along with other domain experts, constitute an informal panel of experts; their responses, combined with other research, formed the basis for our observations.

Suggestions (see pp. 49–70)

We were not able to quantitatively determine the economic benefit, partly because we lacked the market information necessary to assess how the user *values* the services and performance from combined or independent constellations. However, we were able to qualitatively consider the implications for U.S. PNT providers, as well as for users in general. In developing the following list of recommendations, we considered the needs and objectives of the stakeholders (providers and users), as well as plausible civilian user responses to GPS and Galileo coexisting:

1. The United States should remain indifferent to Galileo, from an economic standpoint, as long as the European Union (EU) does not apply restrictive policies/regulations. U.S. responses to such restrictions could include retaliatory practices (e.g., mandating GPS), providing a superior civilian service based on market research, and increasing cooperation with Galileo. We do not rec-

ommend the first action; we view the second and third actions as more likely to result in an increased net economic benefit.

2. The United States should directly address the political impediments to greater cooperation in order to explore the range of options for bringing about greater opportunities in providing PNT data/services. It is important for the United States to establish GPS as a trustworthy and reliable resource for the global community, to leverage opportunities (such as Galileo) to modernize GPS and offer enhanced augmentation services, and, potentially, to maximize GPS's use for future coalition operations. Working with the EU as a cooperative partner in the provision of PNT data/services may help attain these goals.

3. The United States should reevaluate the *implications* of GPS's dual-asset nature. Clearly GPS is and will remain a dual-use system, but a potential opportunity exists to improve the civilian service in ways the United States can do only if it shares the burden. Should the United States seek to formally share the responsibility of satisfying civilian user needs with the EU? Included in this decision is another one: What level of commitment will GPS providers offer to the civilian user base above and beyond what is currently offered? Both the GPS and the planned Galileo system are trying to provide a level of robustness and service that is difficult to meet individually but may be more easily achieved jointly. A combined system may allow both the United States and the EU to provide high performance and robustness without maintaining the current 24+ satellite constellation at all times. This possible scenario—combined, cooperating GPS and Galileo systems—should be examined in earnest but raises many additional questions that require further analysis and evaluation, such as: How much U.S. independence is needed and how much interdependence is tolerable, particularly for national security concerns? What metrics are available for assessing how well these changes would meet U.S. national security objectives, missions, and concerns? What assurances would be required of the EU to demon-

strate its commitment as a reliable partner capable of developing, deploying, and sustaining the Galileo constellation over time? What would be the impact on the many and diverse augmentations that have emerged to satisfy the growing civilian need?

Acknowledgments

We appreciate the many perspectives and suggestions that were provided to us in our examination of a GPS and Galileo global navigation satellite system (GNSS). A broad range of individuals gave of their valuable time to discuss issues related to the GNSS, including Alison Brown, NAVSYS; John Betz, MITRE; Ann Ciganer, Trimble; Eugene Hunt, The Aerospace Corporation; USAF Colonel Rick Reaser, GPS Joint Program Office; Tom Stansell, Stansell Consulting; Ray Swider, OSD; and Dave Turner, IGEB.

We would also like to thank several RAND colleagues for their contributions to the project. USAF Colonel Ed Blasi, a RAND Fellow while this study was being conducted, provided invaluable insight on GPS management organizations and perspectives from outside the United States. Tim Bonds provided insightful comments on the manuscript in process. And we would like to acknowledge our debt to Bob Preston, a key advisor whom, sadly, we lost this year, for his critical input on the study's direction and focus.

Finally, we wish to thank the many representatives from industry (all of whom are listed in Appendix A of this report) who took the time and made the effort to complete and return our survey. Their insights and comments were extremely helpful to our study.

Of course, we take sole responsibility for any errors or omissions in this report.

Abbreviations

ARNS	Aeronautical Radionavigation Service
C/A	Coarse Acquisition Code
CAT	category
COMM	commercial
COSPAS	Cosmicheskaya Sistyema Poiska Avariynich Sudov (Russian for "Space System for the Search of Vessels")
C3I	command, control, communications, and intelligence
DoD	Department of Defense
DOP	dilution of precision
DOS	Department of State
DOT	Department of Transportation
DT&E	development, test, and evaluation
EC	European Commission
EGNOS	European Geostationary Navigation Overlay Service
ESA	European Space Agency
EU	European Union
FAA	Federal Aviation Administration
FOC	full operational capability
FY	fiscal year

GIS	geographic information system
GLONASS	Russian global navigation satellite system
GNSS	global navigation satellite system
GPS	Global Positioning System
GSM	Global System for Mobile communications
ICAO	International Civil Aviation Organization
IGEB	Interagency GPS Executive Board
IOC	initial operational capability
ION	Institute of Navigation
ITRF	International Terrestrial Reference Frame
ITU	International Telecommunication Union
JPO	Joint Program Office
m.a.	masking angle
MEO	medium Earth orbit
MHz	megaHertz
MOA	memorandum of agreement
MSAS	Multi-Functional Satellite Augmentation System
NASA	National Aeronautical and Space Agency
NATO	North Atlantic Treaty Organization
NOAA	National Oceanic and Atmospheric Administration
OASD	Office of the Assistant Secretary of Defense
OS	open service
P(Y)	military code
PNT	positioning, navigation, and timing
PPS	precise positioning service
PRS	public regulated service
RAIM	Receiver Autonomous Integrity Monitoring
R&D	research and development
RDT&E	Research, Development, Test, and Evaluation

RNSS	Radio Navigation Satellite Service
SA	selective availability
SAR	search and rescue
SARSAT	Search and Rescue Satellite-Aided Tracking
SBAS	space-based augmentation system
SIS	signals in space
SoL	safety of life
SPEP	space-based PNT-enabled products
SPS	standard positioning service
SSG-ISC	Senior Steering Group–International Space Cooperation
SV	space vehicle
TAI	International Atomic Time
TCAR	Three Carrier Ambiguity Resolution
TWG	technical working group
UMTS	Universal Mobile Telecommunication System
URE	User Ranging Error
USNO	U.S. Naval Observatory
UTC	Coordinated Universal Time
WAAS	Wide-Area Augmentation System
WGS-84	World Geodetic System—1984
WRC 2000	2000 World Radiocommunication Conference

CHAPTER ONE

Introduction

In December 2001, U.S. Deputy Secretary of Defense Paul Wolfowitz wrote the European Union (EU) officials regarding concerns over "security ramifications for future NATO operations if the European Union proceeds with Galileo satellite navigation services that would overlay the spectrum of GPS (Global Positioning System) military M-code signals"[1] (Wolfowitz, 2001). Additionally, early in March 2002, the U.S. State Department indicated that it would be "unacceptable for Galileo to overlay the same portion of the radio-frequency spectrum used by the GPS military service. The United States would be opposed to anything that would degrade the GPS signals (civil or military), diminish the ability to deny access to positioning signals to adversaries in time of crisis, or undermine NATO cohesion" (U.S. Department of State, Office of the Spokesman, 2002).

Even before Galileo became an official program in March 2002, when its initial funding was made available, the potential advent of Galileo was viewed as a cause for concern. However, this concern was not confined to the technical and national security aspects of the GPS; some viewed the program as a destructive competitive threat.

---

[1] The M-code is an improved navigation signal developed for the military that provides greater protection from interference than the current P-code does.

Background

In 2008, the Europeans plan to begin operating the Galileo positioning, navigation, and timing (PNT) system. In many ways, Galileo is very similar to GPS. Both systems are designed to provide radio navigation signals from a sizable satellite constellation operating in several orbital planes, offering a free service for mass-market applications. In other ways, they are significantly different, particularly in how they are managed and operated. For example, the Galileo business model anticipates generating revenue from fee-based services.

Originally a military warfighting system, GPS is now a widely employed dual-use asset, supporting civilian and commercial users as well as the military. GPS and its U.S. government augmentations are managed by the Interagency GPS Executive Board (IGEB), which was established by Presidential directive in 1996.[2] Figure 1.1 shows the IGEB's composition and its relationship to different agencies. The Department of Defense (DoD) is responsible for acquiring, operating, and sustaining GPS, a responsibility it delegates to the U.S. Air Force. The Department of Transportation (DOT), responsible for acquiring, operating, and sustaining GPS civil augmentations, delegates its responsibility to the Coast Guard and the Federal Aviation Administration (FAA). GPS is a 24/27-satellite constellation[3] operating in six orbital planes at approximately 11,000 miles up. Currently, it transmits separate civilian and military signals at the L1 frequency (1575.42 MHz) and another military signal at the L2 frequency (1227.6 MHz). In the future, new military signals will be available on the L1 and L2 frequencies and new civilian signals will be available on the L2 and L5 frequencies. GPS signals are a free good supplied by the U.S. government as a global utility.

---

[2] http://www.igeb.gov.

[3] GPS's full operational capability (FOC) is defined as 24 satellites with three spares.

Figure 1.1
GPS Management Structure

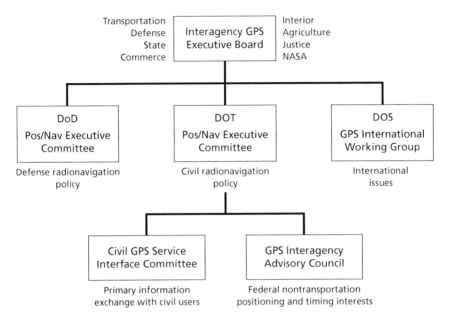

SOURCE: http://www.igeb.gov/org.
RAND MG284-1.1

Galileo, a joint project of the European Commission (EC)[4] and the European Space Agency (ESA), was designed for civil, commercial, and security use but has the potential for use in military applications. The current plan is for Galileo to be administered and controlled by a civilian organization, including a supervisory board representing selected EU countries. Galileo is to be a 27/3/1-satellite constellation operating in three orbital planes at 23,616 km (14,600 miles).[5] It is to broadcast in three frequency ranges—L1 (1,559 to

---

[4] EC is an EU institution with four main roles: propose legislation, administer and implement Community policies, enforce Community law, and negotiate international agreements, mainly those relating to trade and cooperation (http://europa.eu.int/comm/index_en.htm).

[5] Galileo's FOC is defined as 27 operational satellites (nine in each plane), three spare satellites (one in each plane), and one spare on the ground.

1,591 MHz), E5 (1,164 to 1,215 MHz), and E6 (1,260 to 1,300 MHz)—and will offer a basic service, with management seeking partial cost recovery from user equipment royalties and user fees for enhanced services, such as integrity and guarantees[6] (see Figure 1.2).

The Europeans associate Galileo's significance with reduced European dependence on the United States for PNT, the develop-

Figure 1.2
Galileo Public-Private Partnership Overview

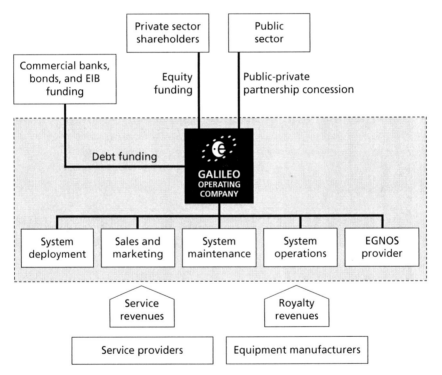

SOURCE: Poulter, 2002.
RAND *MG284-1.2*

---

[6] *Integrity* is the ability to determine whether the system is providing reliable navigation information.

ment of European technical capabilities, and the mitigation of the (current) GPS shortcomings in accuracy, dependable coverage, and notification. However, some in the U.S. Department of State have questioned its need, and others, such as Deputy Defense Secretary Wolfowitz, have indicated that it creates a potential national security issue. This issue stems from the plans to have Galileo overlay a signal on the planned U.S. military signal, and it has both a technical and a geopolitical component. Several groups, including the GPS Joint Program Office (JPO) and NATO, are currently addressing this issue. Therefore, this issue is not addressed in detail in this report.[7]

Talks between the United States and the EU regarding Galileo began in 2000, ostensibly to find a cooperative agreement on how the two systems could provide benefits for users of both systems (U.S. Department of State, Office of the Spokesman, 2002). On several key points, the U.S. government position has remained firm. The March 2002 State Department position (U.S. Department of State, Office of the Spokesman, 2002, p. 2) indicated that

> Europe should not opt to use regulations or system-driven standards to mandate the use of Galileo at the expense of GPS manufacturers, service providers, and users. The U.S. view is that users should be free to choose which system or combination of systems best meets their needs. Similarly, the United States would be against any restrictions on access to information on Galileo that non-European companies may need to participate fully in the equipment and services markets.

On March 26, 2002, the European Council of Transport Ministers released the initial $396 million needed to develop Galileo, officially launching the program.

---

[7] Several news sources have reported that agreement was reached on this issue. For example, according to Peter B. de Selding, in the April 7, 2003, issue of *Space News*, "The European Commission . . . has agreed to U.S. demands that Europe's planned system not use frequencies planned for the future GPS military code" (online at http://www.space.com/spacenews/archive03/galileoarch_041503.html); and *Aerospace America* reported that "agreement had been reached on the issues surrounding potential conflicts with GPS military signals" ("Galileo Takes on GPS," p. 41). At the time of this writing, we were unable to confirm that such an agreement had occurred.

Study Objectives and Sources We Used to Meet Them

The coexistence of GPS and Galileo will create a potentially competitive set of resources for providing PNT data. The extent of that competition will be shaped by many factors, three of which we examine in this study: (1) interoperability and compatibility, (2) strategies employed to foster Galileo adoption, and (3) schedules for GPS modernization and for Galileo development. We sought to identify what economic influence these factors might have on GPS and Galileo coexisting and the implications of that influence for GPS.

Two primary steps were necessary to answer the research questions: (1) define the economic analysis framework, and (2) use that framework to evaluate the economic impact as a result of the competitive factors. Step 2 includes several substeps: (a) characterize the global navigation satellite system (GNSS) by describing the overall architecture as well as the function and performance of the GPS and Galileo components of GNSS, (b) characterize the stakeholders (providers and users of PNT data/services), and (c) estimate user response (and economic benefit) as a function of the competitive factors. We detail each step below.

Step 1: Define the Economic Analysis Framework

Our primary measure of the economic impact of any satellite navigation/timing system, be it GPS only or a combined GPS-Galileo system, is its *net economic benefit* to U.S. citizens, where net economic benefit is equal to gross benefits less costs.[8] The *gross benefits* of the systems are related to the markets for satellite PNT-enabled products—that is, all goods and services that employ satellite PNT information in order to operate and be useful.

The gross benefits of the systems can be divided into three parts:

1. Economic benefits to users of satellite PNT-enabled products/
   services

---

[8] Since both gross benefits and costs occur over time, a proper economic analysis must consider the total future time stream of each, up to some future time horizon (see Appendix D).

2. Economic benefits to providers of satellite PNT-enabled products/ services
3. Economic benefits to the overall economy related to these markets.

We concentrated on the *economic benefits to users of satellite PNT-enabled products/services*, which are defined as the difference between the "willingness-to-pay" of the users of the products/services and the market prices of those products/services. *Willingness-to-pay* is the maximum amount that the product users would be willing to pay for the amount they use, or the users' (consumers') valuation of the products/services provided. The difference between this amount and the total amount paid is called *consumer surplus*, which is interpreted as the economic benefit that the user receives as a result of participating in this market. If there is no change in the consumer surplus as a result of the addition of Galileo, then, in economic terms, there is no benefit (see Appendix D). However, it must be noted that this strict interpretation does not consider the benefits that may accrue to other areas (items 2 and 3, above).

Step 2: Use the Economic Framework to Assess the Influence of the Competitive Factors

Step 2a: Characterize the GNSS. To consider the overall functionality and capability of GPS and Galileo, we used program documentation, various navigation-related studies and reports, and direct interviews with domain experts.

Step 2b: Characterize the Stakeholders. To characterize the stakeholders (providers and users of satellite PNT), we initially interviewed representatives of the GPS JPO, Office of the Assistant Secretary of Defense for C3I, the Institute of Navigation (ION) community, and the GPS Industry Council. Additionally, we held exploratory discussions with several manufacturers and service providers, in which we probed the significance of GPS and Galileo's coexistence and the consequences of the competitive factors. Then, using this information, we developed preliminary observations about the influence of these factors.

Step 2c: Estimate the User Response as a Function of the Competitive Factors. The economic framework attempts to capture the changes in user demand for PNT services/products as a function of the competitive factors. Using the information gathered in Step 2b, we developed a survey (see Appendix B) to further characterize and objectively estimate the possible user response. It included questions related to performance, management, and utilization of satellite PNT information.

The survey was initially tested on a few respondents and then widely and internationally distributed to companies operating in various market segments. We partitioned those market segments as follows: car navigation, consumer/recreational, survey/mapping/GIS (geographic information system), tracking/machine control, aviation, original equipment manufacturing, marine, military and public safety, and timing. Canvassing the PNT industry was the pragmatic approach, because the industry's ability to remain competitive requires an appreciation of the user's needs and constraints. However, of the approximately 250 companies that received the survey, only 7.5 percent completed and returned it.[9] The sample size thus is not representative, so the results are illustrative but not generalizable.

Understanding whether the emergence of Galileo would motivate increased user demand (beyond what would normally occur with GPS), as well as the rationale for any such increase, is a key component of the economic analysis. From the survey and interviews, we learned about the market system performance and management and the utilization of GNSS data and services from the respondent's (i.e., the user's) viewpoint. We were not able to collect enough data to quantitatively analyze the changes in demand and cost relative to the competitive factors, but we were able to qualitatively develop observations about the system capability, system performance, significance of the competitive factors, and implications for GPS.

---

[9] We identified companies using four sources: Fry, 1998; *GPS World Buyers Guide,* 2002; *GPS World Receiver Survey,* 2003; and the GPS Manufacturers Website (http://www.comm-nav.com/gps.htm).

Limitations on the Scope of the Study

The complexity of GPS and Galileo's coexistence necessitated firm boundaries on the scope of this study. As previously stated, the primary objective was to explore the net economic benefit as a result of the three competitive factors.

For the first of these factors, the level of GPS and Galileo interoperability/compatibility, we assessed the economic impact using three parameters: timing, geodesy, and signal structure/frequency (Turner et al., 2002). In this framework, the values for these parameters define conditions that range from *interoperable* (satellite navigation systems are architecturally equivalent, and a single common receiver can use multiple satellite navigation systems) to *compatible* (satellite navigation systems differ architecturally and do not degrade one another, but more-complex receivers are required to use both systems). We bounded our inquiry to consider how these parameters (from the user's perspective) influence the economic benefit.

To interpret the significance of these parameters, we had to consider the designs of GPS and Galileo. However, we neither comment technically on the design of Galileo nor make comparisons with the intention of ranking the two systems.

The second factor is the strategy employed to foster Galileo adoption. The Galileo business model depends partly on revenue from royalties and services that use Galileo (see Figure 1.3). The business case for Galileo is based largely on the projected growth and use of GNSS in various markets. This dependency has raised concern that the EU might employ strategies to ensure that Galileo PNT information is required in some markets and/or that Galileo will be regulated as the PNT provider.

Additionally, motivations related to the development of European technical capabilities, such as increased opportunities for European firms, has raised concern that opportunities for non-European entities may be minimized if regulations restrict non–European company involvement or access to critical information. Therefore, we considered two cases, one in which Galileo is mandated for use in cer-

Figure 1.3
Revenue Generated for an Operating Company

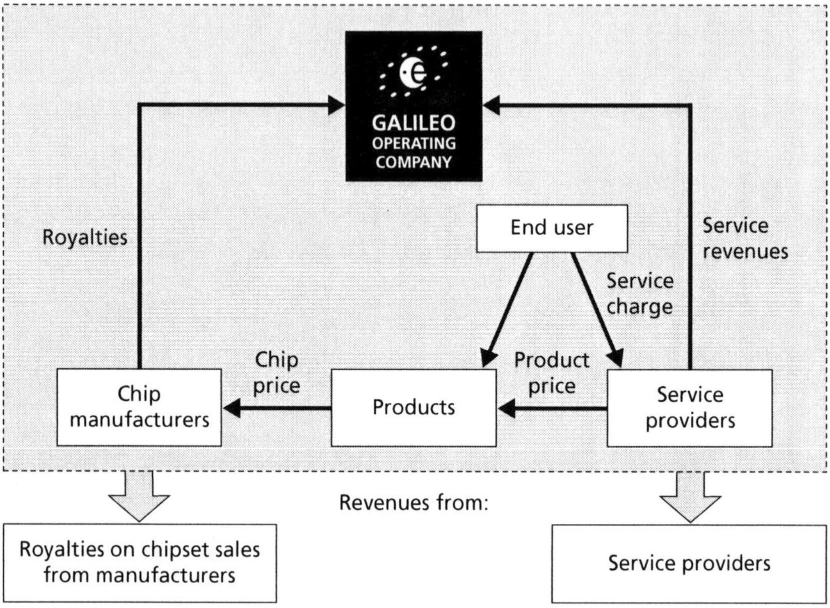

SOURCE: Poulter, 2002.
RAND *MG284-1.3*

tain segments and one in which U.S. companies are inhibited from participating in economic opportunities for Galileo-based services/ products. It was not our objective to determine the soundness of the Galileo business model but, rather, to explore what influence these business strategies, if used, might have on the net economic benefit.

The third factor is the schedules for Galileo development and GPS modernization. Galileo intends to begin operations in 2008 (see Appendix C, Figure C.1 for schedule). GPS will be modernized to include additional civilian services and to improve its performance (see Appendix C, Figures C.2 through C.4). We considered what the economic benefit might be as a result of the incremental capabilities available from GPS and Galileo. We make no comment on the likelihood of either schedule being attained, but we do note challenges facing both efforts.

Clearly, these three factors are not the only ones that will influence the economic impact. In this study, however, it was not possible to identify or consider all the factors of consequence for GPS.

Report Organization

Following this introduction, Chapter Two describes the GNSS and broadly discusses its system segments, GPS and Galileo service offerings, and the system performance of GPS, Galileo, and GPS and Galileo combined. There is also a summary of the user survey responses related to the GNSS's functionality and performance. Chapter Three characterizes the providers of space-based PNT. In addition to detailing their motives and objectives, it explores the challenges and opportunities these providers will face as a result of Galileo. Chapter Four characterizes the users and describes what issues, concerns, and challenges they may encounter as a result of Galileo. A detailed summary of the survey responses is also included. Chapter Five summarizes our analysis of the significance of the competitive factors and their influence on the economic benefit; Chapter Six describes possible market responses to Galileo and offers potential U.S. actions as a function of these market behaviors. Chapter Seven, the final chapter, presents our observations and recommendations.

Four appendices are also provided in this report. Appendix A contains the list of companies with which we interacted, either directly or via survey, for this study; Appendix B contains the survey we used. Appendix C presents the Galileo development and GPS modernization schedules.[10] Appendix D details the economic framework.

---

[10] These were the available and current schedules at the time this study was conducted, in early to mid-2003.

CHAPTER TWO

Characterization of Global Navigation Satellite System

This chapter provides an overview of satellite positioning, navigation, and timing (PNT) architectures. It also compares the services and performance available from the GPS and Galileo components of the GNSS.[1]

System Segments

Space Segment

The space segment consists of on-orbit space vehicles (SVs) that make up the constellation of satellites providing signals-in-space (SIS). The frequency allocations, following the 2000 World Radiocommunication Conference (WRC 2000), for GPS and Galileo SIS are indicated in Figure 2.1.

A nominal GPS operational constellation consists of 24 satellites that orbit Earth every 12 hours. Since a majority of the satellites have performed well beyond their life expectancy, 29 are currently in orbit instead of the designed 27. The constellation has six orbital planes, each nominally with four SVs, that are equally spaced 60 deg apart and inclined at about 55 deg with respect to the equatorial plane. This constellation provides the user with between five and eight SVs

---

[1] Russia's GLONASS is properly a part of the GNSS also, but it was not considered in this study.

Figure 2.1
Spectrum Allocations After WRC 2000

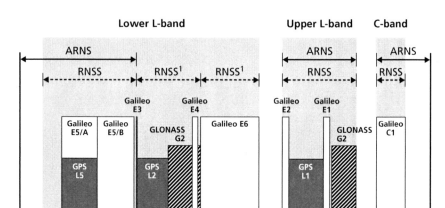

SOURCE: Reaser, 2002.

NOTE: ARNS = Aeronautical Radionavigation Service; RNSS = Radio Navigation Satellite Service.

[1]This RNSS band is shared with other services.

RAND MG284-2.1

visible from any point on the earth. The GPS constellation consists of 21 Block II and Block IIAs, and 8 Block IIRs.[2]

DoD has responded to increased civilian needs by adding new and more-robust civil signals (see Figure 2.2). The second civil signal (L2C) is first implemented in the GPS IIR-M, and the third civil signal (L5) is first present in the GPS IIF. (The schedules for IIR-M and IIF satellites are in Figures C.3 and C.4 of Appendix C.) L2C was

---

[2] The term *block* refers to groups or families of SVs. Block IIs were launched from February 1989 through October 1990, and Block IIA satellites were launched November 1990 through November 1997. The Block IIR satellites are the operational replenishment satellites. They began being launched in January 1997. As of mid-2003 (the end of our study), two IIRs had been launched (January 29 and March 31), and two more were planned for launch later in 2003 (October and December). After these four 2003 IIR launches, ten remaining replacement satellites are to be launched, eight of which are being modernized to carry two new military signals and a second civilian signal.

designed for civilian users who want dual frequency (for ionospheric correction) and for single-frequency users who need better interference protection. Other L2 improvements relative to L1 suggested that L2 "would be the most widely used of all GPS signals" (Fontana, Cheung, and Stansell, 2001).

As of this writing, Galileo plans to broadcast in the L1 and L5 frequencies (along with GPS) but not, as originally intended, in the L2 frequency (Stansell, 2003). Some speculate that receiver manufacturers wanting to use all signals will opt to maximize the cost-effectiveness of their designs by focusing on frequencies that are common across the GNSS, thereby minimizing receiver complexity while taking advantage of a greater number of available signals (Brown, 2003; Stansell, 2003; and our survey responses). This focus on commonality has led to another speculation, that without commonality across the GNSS for L2, commercial and civil markets may provide less support for L2. Further exacerbating the support issue is the fact that the FAA will not use L2, the stated rationale being that the signal is not in the Aeronautical Radionavigation Service (ARNS) band (where civil aviation authorities manage safety-of-life applications) (Stansell, 2003) and that (according to Berger, Lew, and Lane, 1999) it "is not a feasible candidate for safety-of-life applications due to the difficult and high cost of obtaining the required international ARNS spectrum allocation." A reduced civilian base for L2 frequency may make it difficult to protect GPS from encroachment by other users in the coveted L2 band. The opportunity to build a constituency to protect the Radio Navigation Satellite System (RNSS) spectrum is improved if multiple GNSS service providers cooperate to use the same frequencies (Turner et al., 2002).

The Galileo design consists of 30 medium Earth orbit (MEO) satellites, of which three are spares, in a Walker 27/3/1 constellation. The 27 satellites are equally spaced in three orbital planes and have an inclination of 56 deg. The satellites include a navigation and a search-and-rescue payload.

The Galileo satellite system will use ten signals—six for open service (OS) and safety-of-life (SoL) service, two for commercial

Figure 2.2
Planned GPS Signal Structures

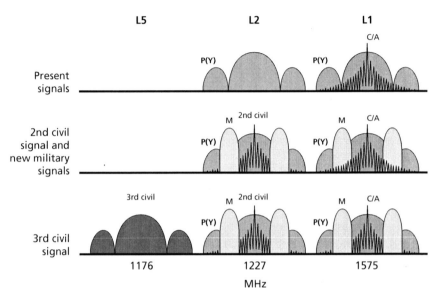

SOURCE: Reaser, 2002.
RAND *MG284-2.2*

(COMM) services, and two for public regulated services (PRS). These will be broadcast in the 1,164–1,215 MHz (E5a–E5b), 1,260–1,300 MHz (E6), and 1,559–1,591 MHz (L1) frequency bands. SoL will be transmitted in the L1 and E5 frequency bands.

Ground (Control) Segment

The ground segment tracks the satellites and transmits corrected orbital parameters. Frequent uploads and corrections from the ground reduce clock errors caused by satellite components and signal errors caused by the atmosphere.

The GPS ground segment consists of a master ground control and four unmanned stations. The Galileo ground station design is based on two ground control centers and a global network of an unspecified number of monitoring stations.

User Segment

The GPS user segment consists of the GPS receivers and the user community. When GPS was first conceived, two signals were provided—a coded signal for the military (P-code) and a noncoded signal for civilians (C/A)—and broadcast on different frequencies. A major source of bias error is the delay of the GPS carrier signals passing through the ionosphere and troposphere. P-code or codeless techniques that measure the phase-delay difference between the code carried on two frequencies (e.g., L1 and L2) are used to reduce the error. For a single-frequency (L1) receiver, the ionospheric delay can be approximated by the single frequency ionospheric model broadcast by each satellite and applied based on user location and time of day.

Augmentations

Space Based or Wide Area. In the United States, FAA and DOT are developing a space-based augmentation system (SBAS) called the Wide-Area Augmentation System (WAAS), primarily for use in aviation.[3] WAAS corrects for GPS signal errors caused by ionospheric disturbances, timing, and satellite orbit errors. WAAS is also intended to provide GPS integrity: vital information regarding the health of each GPS satellite. Another way to provide integrity is the Receiver Autonomous Integrity Monitoring (RAIM) method, which uses algorithms in a receiver to determine integrity. However, RAIM cannot operate without five operational satellites with appropriate geometry, which is beyond the designed capability of GPS (Airways New Zealand, n.d.). The RAIM system is available for civilian use, such as for boaters and recreational GPS users, and it was recently commissioned for instrument flight use.[4]

Other governments are developing similar satellite-based differential systems. The European Geostationary Navigation Overlay Service (EGNOS), an augmentation system designed to complement

---

[3] WAAS currently consists of two Inmarsat-3 satellites. The FAA is trying to acquire a third on-orbit capability for FY 2004 (see http://gps.faa.gov/CapHill/geosat.htm).

[4] See http://www1.faa.gov/index.cfm/apa/1062?id=1756.

GPS and the Russian global navigation satellite system (GLONASS) for civil service, is planned to be operational by 2004. It implements a warning of system malfunction (integrity) within the GPS and GLONASS constellations. In Japan, the Multi-Functional Satellite Augmentation System (MSAS) provides a similar complement. The International Civil Aviation Organization's (ICAO's) international SBAS standards are designed to guarantee the consistency of all these systems at the user level.

Locally Assisted Services. Both GPS and Galileo signals can be enhanced by local augmentations. Differential code corrections from a fixed terrestrial site will reach a nominal positioning accuracy of better than 1 meter. With three civilian frequencies, carrier-phase differential code corrections will enable centimeter-level accuracy (Hatch et al., n.d.), and wide-lane processing, such as the Three Carrier Ambiguity Resolution (TCAR) technique, will enable millimeter-level accuracy (Martin-Neira et al., 2002).

System Services

Table 2.1 describes the services that GPS offers and Galileo plans to offer. GPS now offers two services: (1) a civilian code, the standard positioning service (SPS), and (2) a military code, the precise positioning service (PPS).[5] Galileo intends to offer five services, only one of which, open service (OS), is focused on in this report's analysis.

Table 2.2 identifies the PNT markets used in this study,[6] maps them to the services offered by GPS, and provides a plausible mapping of the Galileo services (except for search and rescue [SAR]) that might be voluntarily used by each market.

---

[5] Even when GPS modernization makes new signals and frequencies available, signal-structure design will still provide the basic separation of civilian and military services: L1, L2, and L5 will carry civil codes while L1 and L2 carry military codes.

[6] The markets for the study also included original equipment manufacturing (OEM) but have been omitted from this table because OEM products are inherently included in the listed applications.

Table 2.1
GPS and Galileo Services

Service	Description	Target	Notes
GPS			
Standard positioning service (SPS)	Free navigation and timing service for use by anyone	Peaceful commercial and civilian uses	
Precise positioning service (PPS)	Encrypted navigation and timing service for use by U.S. military and NATO	Strategic and national security matters	
Galileo (planned)			
Open service (OS)	Free navigation and timing service for use by anyone	Mass-market applications, particularly low-cost/simple receivers, designed to facilitate the use of combined services	No service guarantee, liability, or integrity information available
Safety-of-life (SoL)	Frequency receivers that operate in L1 and E5 (SoL Radio Navigation Satellite Service [RNSS] band)	Transportation applications in which lives are endangered without real-time notice of performance degradation	Service guarantee for SoL applications, integrity notification, signal authentication
Public regulated service (PRS)	Encrypted fee-based navigation and timing servlce for pollce, flre, ambulance, military, and customs; access controlled by civil authorities	Public or strategic applications	Must be operational at all times under all circumstances, and must mitigate interference broadcast on separate frequencies with respect to OS to prevent signal loss when OS is denied
Commercial (COMM)	Encrypted fee-based navigation and timing service for market applications; controlled access via receiver; low data-rate broadcasting capability	Service providers offering applications dependent on this performance	Service guarantee for fee

Table 2.1—continued

Service	Description	Target	Notes
Search and rescue (SAR)	Relays distress alarms to improve existing relief and rescue services	Existing systems for international cooperative effort for humanitarian search and rescue (compatible with Russian COSPAS [space system for the search of vessels] and SARSAT [Search and Rescue Satellite-Aided Tracking])	

SOURCE: Mastracci, 2002.

Table 2.2
PNT Markets Mapped to Services

Market	Service					
	SPS	PPS	OS	SoL	PRS	COMM
Car navigation	Yes	No	Yes	No	No	Unknown
Consumer/recreational	Yes	No	Yes	No	No	Unknown
Survey/mapping/GIS	Yes	No	Yes	No	No	Unknown
Tracking/machine control	Yes	No	Yes	No	No	Unknown
Aviation	Yes	No	Yes	Possibly	No	Unknown
Marine	Yes	No	Yes	Possibly	No	Unknown
Military and public safety	Yes	Yes	Yes	No	Yes	Unknown
Timing	Yes	No	Yes	No	No	Unknown

In terms of functionality and potential market application, the OS is comparable to the SPS, and the PRS is comparable to the PPS. New civil capabilities that Galileo will provide and that GPS has not stated any intention of providing are the service guarantees (certification/liability) and communication functions. Civil users will be able to receive integrity data through the fee-based services but not through OS. Integrity is accomplished for GPS users via application of augmentation systems.

System Performance

The performance that a user experiences (in terms of accuracy, availability, etc.) is the result not only of conditions such as constellation geometry and signal power, but of other factors as well, including control segment actions, receiver equipment capability or configuration, use of augmentation, and interference.[7] We asked the survey respondents to indicate the criticality of these parameters and their need for improved parameter performance/service in their market/business. Our survey included questions related to the performance of the seven parameters listed in Table 2.3.

Criticality of Performance Parameters Survey Response
Table 2.4 shows the responses of the survey respondents when asked to indicate the criticality of the parameters.

GPS, Galileo, and GPS-Galileo Performance
Tables 2.5, 2.6, and 2.7 set forth the existing or anticipated performance value for civilian/OS services from, respectively, GPS, Galileo, and the GPS-Galileo combination.

The varied position accuracy values reflect the fact that performance depends on the receiver equipment configuration. The number of "satellites-in-view" will change as a result of masking angle (m.a.), and the effectiveness of ionospheric correction will differ as a result of whether a single or a dual frequency is used.

Satellite-only navigation systems, such as GPS, the future Galileo, or even a combined GPS and Galileo, cannot meet some stringent requirements. For example, GPS and Galileo, whether alone or

---

[7] Interference can be intentional or unintentional. *Intentional* interference may be associated with a crisis or criminal activities. *Unintentional* interference may occur for many reasons, from malfunctioning equipment to uncontrolled/unmonitored use of spectrum. For instance, concerns are mounting about potential interference between existing radio-communications systems and emerging ultra-wide-band (UWB) technology emissions in Europe (particularly below 3.1 GHz). The concern is that the use of commercial UWB devices may raise the noise floor, which will be a problem for GPS receivers in Europe. (Enge, 2002.)

Table 2.3
Seven Parameters Related to Performance and Their Definitions

Parameter	Definition
Position accuracy	Statistical value of error between true position and estimated position; measured as a distance (e.g., horizontal distance and vertical distance) at a stated confidence level
Availability	Percentage of time that position accuracy meets specified accuracy performance level
Continuity gap	Maximum continuous length of time that specified position accuracy not met without advance notification
Integrity	Ability to determine whether system is providing reliable navigation information; measured as rate at which system will not provide user with hazardously misleading information (e.g., $X * 10^{-Y}$/second)
Time-to-alarm	Length of time required to provide notification at user interface that service is unavailable
Timing accuracy	Statistical value of error between true time (Coordinated Universal Time [UTC]) and estimated time; current specifications state that at 95% confidence level, timing accuracy shall be no greater than 20 nsec for static user and 35 nsec for dynamic user
Guarantee	Concept of ensuring services for applications in which a disruption of service would have significant SoL or economic effects

SOURCE: *GPS ORD'99.*

Table 2.4
Survey Respondent Assessments of Performance Parameters (counts)

Assessment	Performance Parameter						
	Position Accuracy	Avail-ability	Continuity Gap	Integrity	Time-to-Alarm	Timing Accuracy	Guarantee
Essential	8	13	9	10	4	8	8
Somewhat important	9	4	4	6	10	4	5
Not important	0	0	3	1	3	5	4
No response	2	2	3	2	2	2	2

Table 2.5
GPS Performance Specification

Position accuracy[a]	Current specifications state that at 95% confidence level, position accuracy shall be no greater than 4.0 m horizontal and 7.6 m vertical
Availability	Dependent on constellation status
Continuity gap	N/A
Integrity	N/A (augmentations are used to provide integrity)
Time-to-alarm	N/A (this capability included in WAAS)
Timing accuracy	Current specifications state that at 95% confidence level, timing accuracy shall be no greater than 20 nsec for static users and 35 nsec for dynamic users
Guarantee	N/A

SOURCE: *GPS ORD'99*.

NOTE: N/A = not available.

[a] The position accuracy of GPS is based on both the User Ranging Error (URE) and the satellite geometry, which causes dilution of precision (DOP). URE is an error vector along the line of sight (LOS) between the user and the satellite. It consists of the signals-in-space (SIS) error (which is due to satellite clock anomalies, errors in the broadcast navigation message, atmospheric delay, and multipath) and the user equipment error. DOP is a measure of how satellite geometry affects accuracy.

combined, cannot satisfy the FAA requirement for landing (CAT II/III), for which the vertical alarm limit (VAL) cannot exceed 4 to 10 meters. Space-based augmentation systems (SBASs) have been developed to provide more-stringent services. Table 2.8 illustrates the performance level of GPS, Galileo, and GPS-Galileo, alone and with SBASs.

Improved Performance Needs Survey Response
We asked the survey respondents to indicate their need for improved performance/service for their market/business according to the seven parameters in Table 2.4. Their responses show that the relationship between system performance and the GNSS market is quite complex. Multiple survey respondents specified that in addition to the essential nature of all seven defined parameters, they needed improvements in each of these parameters for their business/product. They identified position accuracy and availability in particular. The market segment for timing also expressed the desire for enhanced timing accuracy.

Table 2.6
Galileo Performance Specification

	Open Service (Positioning)		Safety-of-Life Service		Public-Regulated Service
Receiver type — Carriers	Single frequency	Dual frequency	Three frequencies		Dual frequency
Computes integrity	No		Yes		Yes
Ionospheric correction	Based on simple model	Based on dual-frequency measurements	Based on dual-frequency measurements		Based on dual-frequency measurements
Coverage	Global		Global		Global
Accuracy (95%)	H = 15 m, V = 35 m	H = 4 m, V = 8 m	Critical level: H = 4 m, V = 8 m	Noncritical level: H = 22 m	H = 6.5 m, V = 12 m
Integrity — Alarm limit	N/A		H = 12 m, V = 20 m	H = 446 m	H = 20 m, V = 35 m
Time-to-alarm	N/A		6 sec	10 sec	10 sec
Integrity risk	N/A		3.5×10^{-7}/150 sec	10^{-7}/hr	3.5×10^{-7}/150 sec
Continuity risk			10^{-5}/15 sec	$10^{-4} - 10^{-8}$/hr	10^{-2}/15 sec
Certification/liability			Yes		
Timing accuracy with respect to UTC/TAI					100 nsec
Availability			98%		99.5%
Availability of integrity	99.5%				
Availability of accuracy	99.8%				

SOURCE: *Galileo Services and Architecture*, 2002.

NOTE: H = horizontal; V = vertical; N/A = not available.

Table 2.7
GPS-Galileo Combined Performance

	10-deg Masking Angle, Single-Frequency Receiver		10-deg Masking Angle, Dual-Frequency Receiver		30-deg Masking Angle, Single-Frequency Receiver	
	Galileo OS	Galileo OS + GPS	Galileo OS	Galileo OS + GPS	Galileo OS	Galileo OS + GPS
Horizontal accuracy (m)	15	7–11	4	3–4	14–54	11–21
Vertical accuracy (m)	35	13–26	8	6–8	21–81	17–32
Availability	99%, worldwide					

SOURCE: *Galileo Services and Architecture,* 2002.

One interesting point was an apparent segmentation of the market between respondents who already make use of augmentation (i.e., a differential GPS network) and those who do not: Those not currently using it appear to prefer enhanced performance, especially in position accuracy, timing accuracy, and availability.

What's New and Improved?

The *new* capabilities (from satellite PNT systems) that are to result from Galileo are service guarantees and low-data-rate broadcasting. Integrity and time-to-alarm are not capabilities of GPS proper, but of GPS combined with WAAS (or other augmentation/integrity-monitoring systems). The performance of the remaining parameters we considered is comparable, suggesting that Galileo by itself does not offer significant performance improvements over GPS with augmentations.

However, GPS and Galileo combined could significantly increase the total number of available satellites in the sky and thereby lead to the following performance improvements:

Table 2.8
Satellite PNT Performance Versus FAA Requirements

Scenario		1	2	3	4	5	6
Architecture	Requirements	Galileo L1/E5	Galileo L1/E5 + GPS L1	Galileo L1/E5 + GPS L1 + SBAS L1	GPS L1/L5 + SBAS L1/L5	Galileo L1/E5 + GPS L1/L5	Galileo L1/E5 + GPS L1/L5 + SBAS L1/L5
Operation	VAL						
Oceanic	N/A	✓	✓	✓	✓	✓	✓
En route	N/A	✓	✓	✓	✓	✓	✓
Terminal	N/A	✓	✓	✓	✓	✓	✓
NPA	N/A	✓	✓	✓	✓	✓	✓
APV I	50 m	✓	✓	✓	✓	✓	✓
APV II	20 m	✓	✓	✓	✓	✓	✓
CAT I	12 m	X	X	✓	✓	X	✓
CAT II/III	4–10 m	X	X	X	X	X	X

SOURCE: Bruns, 2002.

NOTE: VAL = vertical alarm unit; N/A = not available; NPA = nonprecision approach; APV = approach with vertical guidance.

- *Availability:* Improved visibility and robust solutions for applications requiring redundancy for safety and/or security reasons.
- *Position accuracy:* Better geometry of visible satellites and thus reduced DOPs, which means enhanced positioning performance. One drawback here is that having more satellites can lead to more noise, which is not good for interference reduction.
- *Integrity:* Improved integrity performance for SoL applications, such as those required by the FAA; and for an improved RAIM capability.

The potential for a combined system to provide these improvements is tied to the degree and nature of cooperation and competition between the two systems. The competitive factors of this study were specifically selected to explore these issues. Cooperation and competition are as much technical issues as they are geopolitical issues.

Those entities likely to be affected by Galileo can be broadly grouped into GPS providers, Galileo providers, and users. In the next two chapters, we examine the stakeholders—providers and users—who shape the geopolitical landscape. We explore their motivations, concerns, and opportunities in both a cooperative and a competitive environment between GPS and Galileo.

CHAPTER THREE

Providers: Satellite Positioning, Navigation, and Timing

This chapter characterizes the providers of satellite PNT. In addition to detailing their motives and objectives, we explore the challenges and opportunities they will face because of Galileo.

GPS

Those responsible for providing GPS-related capabilities are represented at the Interagency GPS Executive Board (IGEB), a policy-making body established in 1996 by Presidential Decision Directive NSTC-6. The organization is co-chaired by the Department of Defense and the Department of Transportation but includes other agencies and departments as well (see Figure 1.1 in Chapter One). The IGEB manages and operates GPS and its augmentations, consistent with the following guidelines:

- Provide SPS for peaceful civil uses continually, globally, and free of charge.
- Provide uninterrupted service for civil uses; however, reserve the capability to deny GPS services to adversaries in areas of crisis.
- Consider international civil, commercial, scientific, and security needs.
- Promote GPS and its augmentations as the global standard.
- Purchase commercially available GPS products when possible.

In our study, we considered the objectives and needs of the departments of Defense, State, Commerce, and Transportation in answering the following question: Given the roles of these departments in providing GPS capabilities and service, what are their motivations, concerns, and opportunities in a cooperative and a competitive environment between GPS and Galileo?

Department of Defense (DoD)

DoD's role as provider could be further categorized as that of operator and developer.[1] As the operator, the military is responsible for day-to-day maintenance and control of the constellation, as well as for providing support functions to users, such as advisories and notices, as necessary. As the developer, DoD (specifically, the Air Force) is chiefly responsible for funding and executing the program plan that the IGEB sets for GPS.

From a military perspective, the tasks of maintaining and enhancing GPS to meet current and projected needs require that operational concepts be taken into account. One of the more challenging of these concepts since Galileo's emergence is the fundamental intent of one of the objectives from Presidential Decision Directive NSTC-6: "Develop measures to prevent the hostile use of GPS and its augmentations to ensure that the United States retains a military advantage without unduly disrupting or degrading civilian uses."[2] This concept is the centerpiece of the national security–related concern raised by U.S. officials, a concern we refer to in various parts of this report but that was not part of our study's scope. What is noteworthy about the concept is its acknowledgment that an adversary may be able to use open and free satellite-based PNT data with sufficient performance that the United States will be compelled to deny those data to the adversary.

---

[1] DoD is both a provider and a user (as are other groups). In this report, we focus on its role as provider, since its role as user involves national security issues that were beyond the scope of our study.

[2] Quoted from p. 4 of NSTC-6, which was issued by the White House on March 28, 1996. Full text of this directive is online at www.fas.org/spp/military/docops/national/gps.htm.

Additionally, other operational activities and recent engage-ments suggest at least two trends in GPS use: The use of GPS in sys-tems and weaponry is approaching ubiquity, and military operations tend to include coalitions of countries. Is ubiquitous use of satellite PNT data better served by a cooperative or a competitive set of sys-tems? Does increased coalition activity suggest motivations for shared or cooperative investments in space technologies? If GPS and Galileo evolve to greater interdependence, what is the potential for greater efficiency and other improvements (e.g., better performance, shared control/monitor assets)? With greater interdependence, the United States may realize opportunities to alter its investment in GPS and to share the responsibility for growing civilian needs. The extent to which the investment could be altered would, of course, be bounded by the architecture needed to retain U.S. military objectives (Gholz, 2002).

Department of State (State)
The Department of State is responsible for coordinating and/or con-sulting with foreign governments and international organizations re-garding the provision and use of GPS services and augmentations. Such coordination includes negotiating international agreements and controlling the release of U.S. technology under the International Traffic in Arms Regulation and Arms Export Control Act. The De-partment of State's Office of Space and Advanced Technology Staff is responsible for handling/resolving international space issues and sci-ence and advanced technology questions, including GPS-related is-sues, for the Department. Its goals are to ensure that U.S. space-related policies support U.S. foreign policy objectives and to ensure that U.S. international initiatives and political commitments on space are science based, protect national security, advance economic inter-ests, foster environmental protection, and enhance U.S. space leader-ship and the competitiveness of the U.S. aerospace industry.

The goal of enhancing U.S. space leadership and aerospace in-dustry competitiveness appears to be incompatible with efforts to fos-ter greater cooperation with the EU, which has its own goals of lead-ership and competitiveness for the EU countries. The State

Department has expressed concerns about (1) European regulations and standards that may effectively mandate use of Galileo, (2) opportunities to provide navigation products not being made equally available to all manufacturers, and (3) Galileo's strategic and military implications.

Department of Commerce (Commerce)

The Commerce Department is the largest GPS constituency—it includes commercial industry, commercial end users, the National Geodetic Survey, the National Weather Service, National Oceanic and Atmospheric Administration (NOAA) Corps, the National Institute of Standards and Technology, the Census Bureau, and the GPS Interagency Advisory Committee (U.S. Department of Commerce, 2002). The Department manages the federal radio spectrum and supports efforts to protect the GPS radio frequencies; it also hosts the IGEB Executive Secretariat, with the Staff Director also operating as Director of the Office of Space Commercialization. The Office of Space Commercialization, NOAA, and the National Telecommunications and Information Administration are the primary focal points for GPS-related issues within the Commerce Department.

The Office of Space Commercialization is part of the U.S. delegation negotiating with the EU on future GPS and Galileo cooperation. In this role, the Office represents the interests of the commercial GPS community—that is, service and equipment suppliers, as well as consumers. The primary mission for the Commerce Department is to promote commercial market growth and trade. Naturally, areas of concern about Galileo are open market access for U.S. providers of GPS equipment and services, equal access by all parties to future Galileo signals and technical material essential for manufacturing compatible equipment and services, interoperability, and protection of current GPS services from interference. Successful negotiations in these areas could result in lower costs for consumers through free-market competition and a level playing field. Note that the Office of Space Commercialization is co-chair of the working group on commercial and scientific cooperation with Japan on the future Japanese GPS.

Department of Transportation (DOT)

DOT has dual roles as a developer and a user of GPS augmentations. As a developer, it is responsible for consolidating federal civil issues, implementing and coordinating U.S. government civil augmentations for transportation applications, and working with the Commerce and State departments to promote GPS commercial applications and standardization. Through the FAA, DOT is developing the WAAS; through the Coast Guard, it is developing a local area augmentation system.

DOT has many of the same concerns as the other IGEB members, as well as a concern about EGNOS. Currently used to augment GPS in Europe, EGNOS may be recapitalized as the initial satellites of the Galileo constellation. DOT and others are concerned about whether EGNOS would continue to provide GPS data, which are vital to aviation and other users in Europe, if this were to happen. (U.S. Department of State, Office of the Spokesman, 2002.)

Galileo

ESA and EC have partnered to establish the Galileo satellite navigation system. A new agency, called the Galileo Joint Undertaking, will manage the Galileo project. Many of the project's objectives likely gave rise to the need for Galileo, and two primary objectives are currently sustaining support for Galileo's development: to mitigate current GPS shortcomings and to reduce European dependence on the United States for satellite PNT.

Mitigation of Current GPS Shortcomings

Before authorizing Galileo, the EU expressed interest in GPS's operations and control, particularly GPS's ability to provide integrity monitoring and liability guarantees. Integrity monitoring was seen as important for determining how much GPS could be relied upon and how much effort above and beyond GPS would be required if it could not be considered trustworthy. As for the liability guarantees, the EU considered those to be just as important for commercial business needs (Loverro, 2002). How satisfactorily these inquiries were answered is unclear; what is clear is that the EU concluded that GPS's performance in accuracy, dependable coverage, and notification was less than sufficient. Accuracy and notification (immediate warning about signal interruptions and signal errors) are areas that the managers of GPS and/or its augmentations intend to improve.

The EU's expressed challenges with respect to dependable coverage—position (high latitudes), dense areas, and military priorities—have not been directly addressed by the planned GPS enhancements, however. This may have contributed to the EU's decision and stated objective that Europe should be less dependent on the United States for PNT and should develop its own system to improve the EU space industrial base.

Reduction of Dependence on the United States for Satellite PNT

The GPS, although a dual-use system, is nonetheless a military system. From the European perspective, dependence on this U.S. system limits military and industrial independence and is inconsistent with sovereignty. Given that most new weapon systems and platforms incorporate satellite navigation technology, an independent system would improve the European defense posture.

Industrially, satellite PNT systems have wide civilian application and support key infrastructures, such as telecommunications, transport, and transaction networks. With so much of the EU's economic activity based on GPS, continued reliance on a U.S.-controlled system might seem imprudent to European leaders. What if the United States decided to restrict or change access to GPS signals? What if at some time in the future the United States decided to no longer depend on GPS as it does now and found a new service?[3] (Loverro, 2002.)

---

[3] This possibility has been openly expressed. Owen Wormser, the Pentagon's principal deputy for spectrum, space sensors, and command, control, and communications (C3), was quoted in *Aviation Week & Space Technology* as saying "the military also should start thinking about whether it wants to pursue alternative approaches to providing precision navigation and timing data" (Wall and Covault, 2002, p. 31).

From the EU's perspective, an alternative that counters the independent U.S. control of GPS not only would address these challenges, but also could be used to develop European technical capabilities and create economic opportunities. The EU anticipates that the Galileo program will create new jobs in Europe. It also hopes to preserve or create market position for European firms via competition with the United States for alternative commercial application standards and armament sales for precision-guided weapons (Mastracci, 2002). In addition to these goals, Europe believes it can structure Galileo to help offset its costs and support other business (Sabathier and Sapolsky, 2002; Gholz, 2002).

Sovereignty and autonomy motivated the initiative to develop a European alternative. However, the Galileo system design is not totally independent of GPS. The two systems share frequencies, which is advantageous for users because it facilitates their use of both systems, but is a disadvantage in terms of truly independent system management because of user expectations/needs where the systems intersect. Furthermore, some of the signals planned for Galileo directly counter U.S. desires for separation of military signals, raising questions about the motivation for Galileo.[4]

Cooperation or Competition?

Chapter Two concludes with a call for greater U.S./EU cooperation. Opportunities for shared or cooperative investment require greater dependence between the systems and, by extension, greater coordination of the policies and objectives of those reliant on the systems. But the obstacles that must be overcome on this route are nontrivial.

---

[4] The M-code-overlay issue is an exception whereby Europe seems to have defined spectrum codependence for security services as part of Galileo's design. The EU expects the United States to trust that the EU can secure Galileo's PRS and that commonality of U.S. and EU security interests would rule out the need for the United States to deny PRS in any area. The U.S. (and NATO) position is that the United States must be able to deny an adversary access to any GNSS other than M-code in a region during conflict and yet retain its own security use of GPS through M-code.

First, the necessary level of cooperation may be just too difficult to attain and there may be too little support for it. Second, Europe and the United States would have to compromise on some of their overall objectives. Third, U.S. support for the Galileo initiative has only recently materialized.

The needed scope of cooperation appears to be orders of magnitude above and beyond the current minimalist approach both sides are taking. Current discussions for agreements and workarounds to mitigate the *differences* between the systems focus on the user level, where cooperation means only that the systems do no harm to each other. Given that reaching even these agreements, which are as much political as they are technical, was a Herculean task, the likelihood for greater cooperation appears remote. Both the United States and Europe seem to have adopted the position that GNSS is so important to sovereignty that there can be no external dependence. This is an extreme position, one that not only limits opportunities for efficiency, but also may encourage others to follow suit, which could lead to a proliferation of GNSS standards that will make it more difficult for both the United States and the EU to sustain their objectives. And there are other, more direct impediments for cooperation, such as current U.S. arms export control laws that restrict cooperation with foreign governments, particularly when the technology may be transferred to a company.

Europe and the United States have opposing visions for Galileo and GPS. Europe, recognizing its large and growing reliance on satellite PNT data, envisions a greater role and voice for itself in the control of this information utility. The United States, seeking to promote commercial applications of GPS technologies and forge agreements with foreign governments and international organizations, envisions a world where no near-peer global satellite PNT system exists. If GPS and Galileo are to move toward greater interdependence, these visions must be adjusted. Without compromise it will be difficult to achieve greater cooperation and, equally as important, easier to interpret actions as competitive.

Even though the United States and Europe have been discussing Galileo for several years now, the U.S. reception of Galileo only began to warm as the European plans for Galileo evolved. Until the Galileo development program became a reality in 2002, GPS was the only practical option for global satellite-based PNT information. There was no threat of competition—and certainly no need for cooperation.

We were unable to precisely recreate the interaction between the United States and Europe during the pre–March 2002 period, but what our research made clear is that the Europeans at some point decided that they had to develop and control a system independent of the United States (Loverro, 2002). It is not clear whether the United States viewed Galileo as a cooperative opportunity or a competitive challenge once the initiative began to gather momentum, but the official U.S. position was clearly that a second global satellite navigation system was unnecessary. Since March 2002, the United States and Europe have increasingly been interacting in a cooperative manner but subject to the constraints we have been discussing here.

In the next chapter, we consider Galileo's implications for the other set of key stakeholders—the users.

CHAPTER FOUR

Users: Satellite Positioning, Navigation, and Timing

This chapter describes the use and applications of satellite PNT data, including the changes in usage that have occurred as a result of GPS actions, the markets that use satellite PNT data, and the planned objectives for applying satellite PNT data. We used the responses from industry to characterize the user and to consider the challenges and opportunities specific to the user. This characterization includes the users' plans for the use of satellite PNT data, perspectives on operating in a global environment with multiple GNSS elements, and recommendations for facilitating progress in the use of satellite PNT data.

GPS Use and Applications

Beginning with the availability of GPS in 1993 for commercial and civil use, and including the discontinuance of selective availability (SA)[1] in 2000, satellite PNT data have been increasingly offered to users in a variety of applications to improve their living conditions and productivity. A 1998 International Telecommunication Union (ITU) report noted two emerging trends in the GPS market that have further fostered widespread use and adoption of navigation and timing data. The first trend was the declining cost of the receiver: "a con-

---

[1] This is a technique that the United States had used to intentionally degrade the accuracy of the navigation solution.

tinuous 30 percent per year decline in the cost (which is being driven towards zero), power and size of the electronics hardware necessary to decode the space-based information" (International Telecommunication Union, 1998, p. 1). The second trend had two components: the increased presence of software to provide enhanced value for end-user applications and increased user demand, which drove shorter product cycles.[2]

GPS usage is diverse and undoubtedly increasing. The 1998 ITU report identifies 159 civil, commercial, and consumer applications of GPS, the majority of which are in environmental protection, public and ground transportation, infrastructure, aviation, marine, and public safety. The GPS Applications Exchange[3] collects and maintains examples of GPS applications from around the world. As of this writing, it lists approximately 270 uses of GPS, the top spots going to public health and safety, mining/construction, infrastructure, forestry/agriculture, marine, transportation, and surveying/mapping. Based on revenue, car navigation and consumer applications are the leading and fastest-growing applications (International Telecommunication Union, 1998; Frost & Sullivan, 2000).

The total user market is too large and diverse for meaningful analysis (the challenges are described in Appendix D), so we decomposed the market into segments consistent with current applications/use, the industrial base, and existing market data. We asked the GPS industry (our user proxy) to use the following segmentation when completing the survey for our study: (1) car navigation, (2) consumer/recreational, (3) survey/mapping/GIS, (4) tracking, ma-

---

[2] According to the 1998 ITU report (pp. 1–2): "The first commercial receivers cost over $150,000 and weighed over a hundred pounds. The next year a portable receiver was introduced that weighed only 40 pounds and cost $40,000. In 1998, a consumer handheld receiver costs $100 and weighs 7 ounces. . . . The second trend is the increased contribution of embedded software in the end-user application. User demand is driving product evolution with product cycles of 12–18 months. The value to the end-user is in the application of the information, which is recovered largely in software. Now, $100 buys a 3.1-oz receiver: e.g., http://shop3.outpost.com/product/3545584."

[3] http://gpshome.ssc.nasa.gov/appinfo.asp.

chine control, (5) aviation, (6) original equipment manufacturing, (7) marine, (8) military and public safety, and (9) timing.

Of the survey responses, 50 percent came from non-U.S. companies (see Appendix A for a list of the respondents). Table 4.1 describes the survey respondents in terms of business type, market segments served, and notable information about their applications/use.

User Plans for Satellite Navigation Information

Most respondents indicated that they were actively pursuing options for increased performance, because they expected performance improvements to bring new business opportunities and growth—and, in some cases, lower cost. Other respondents were pursuing performance improvements as a means of marketplace preservation. Most respondents chose not to share their approaches for improving performance,[4] but we did learn that strategies such as terrestrially based navigation (e.g., cell sites of telecommunications networks) were being considered.

Many in the GPS community expect that improvements in timing accuracy, precision, availability, and reliability will enhance the functionality of existing products and create new applications, thereby opening new markets and increasing the market adoption of products and services.[5] Some respondents indicated that such performance improvements as service guarantees would enable them to reduce their dependence on more-costly alternatives (such as high-

---

[4] Reasons cited for not sharing this information were that it was company policy not to do so and that sensitive information regarding future markets, plans, and products needed to be protected.

[5] For example, Dr. Gerard Lachapelle, ION Western Region Vice President and head of the Geomatics Engineering Department at the University of Calgary, suggested that the market for personal indoor users (i.e., cellular phones and personal digital assistants) will benefit very significantly from second- and third-generation GPS and the addition of the EU's Galileo, and that 30 to 40 percent of the applications based on satellite navigation are yet to be discovered but will place more and more demands on performance (Lachapelle, 2003).

Table 4.1
Survey Respondent Business Market/Applications

Business Type	Market Segment[a]	Application/Use Information
Manufacturer/supplier of GPS receivers and related products	1–6, 8, 9	• Users requiring 5 to 20 cm accuracy real-time (or better, for some markets) • High-precision (cm positions) • Positioning, velocity and time, radio-communication synchronization and integrated navigation/communication systems • Software algorithms and digital signal processing • Equipment for mandatory E911 requirements in U.S.
Provider of GPS satellite-correction service	1–4, 6, 7	• Real-time decimeter accuracy worldwide • 10 cm 1-sigma worldwide coverage
Manufacturer/supplier of precise time and frequency equipment	8, 9	• Equipment using GPS as primary reference • Discipline of internal clock of a GNSS receiver for timekeeping, timing, location, positioning, and network synchronization applications • Synchronization of timing signals and discipline of oscillators for high-stability frequency reference applications • Telecom network synchronization and test and measurement
Research, Development, Test, and Evaluation (RDT&E)	1, 2, 8	• Software development and equipment testing • GPS constellation performance assessment
Service providers: location-based applications	3, 4	• Consolidation of field-collected/field-mapped data • PPS signal for synchronization purposes

[a] 1 = car navigation, 2 = consumer/recreational, 3 = survey/mapping/GIS, 4 = tracking/machine control, 5 = aviation, 6 = original equipment manufacturing, 7 = marine, 8 = military and public safety, and 9 = timing.

stability oscillators for timing applications). Others indicated that they expect greater availability of the satellite navigation signal to lower component costs, such as those for avionics.

Still others are looking to improve performance as a means of protecting or preserving their business base. The objective of augmentation systems is to deliver performance above and beyond what is otherwise achievable. As base performance improves, augmentation systems are compelled to further improve or find some other value-added service. Respondents with this concern suggested that if they do not offer a better service, their market share will initially remain constant but then begin to decline. For systems whose business market is based on providing performance enhancements related to the GPS (which includes the FAA's WAAS), Galileo may appear to be a threat.

These responses clearly indicate that the market is looking for improved performance—to develop new markets in some cases and to sustain current market advantage in others. We also see clear evidence for the use of augmentation systems or other alternatives as a way to provide improved performance. Given that GPS and Galileo will offer fairly comparable performance, we expect this same type of performance-seeking behavior to continue, regardless of which system the user relies on.

For a user able to use both systems, this may obviate augmentation systems or other alternatives. What is not clear, and what we were unable to distinguish, is whether improvements in performance as a result of GPS and Galileo being combined will drive a significant increase in demand, serve only to maintain current growth, or have no effect at all because the market is already maturing and saturated.

Operating in a Global and Multiple-System GNSS Environment

Of the many uncertainties regarding the future GNSS, the possibility of regulation and the effect of civilian (as opposed to government/military) management and control garner much attention because of their potential for influencing what would otherwise be open-market competition.

The respondents that addressed the issue of regulation—mandated use or limited/restricted business opportunities—did not agree about the impact of these actions, and their differing perspectives tended to depend on their market segments. Those in the business of manufacturing or supplying GPS equipment (i.e., receivers) were more sensitive to mandated use—that is, they saw it as more negative than others did. Their concern was the additional R&D required to support each component of the GNSS, whereby cooperation would be sacrificed for competition. One survey respondent wrote: "A non-generic solution will have a large impact that affects design cycles, customer training, etc." For those companies that market internationally or have customers who need worldwide solutions, regional differences would be problematic, necessitating strategies to mitigate various environments as a result of competition. This was not the view of certain other respondents, however. Those that provide RDT&E or software solutions, for example, saw mandated use as a positive because of the potential for new or expanded business opportunities. And some respondents were indifferent, believing that there would be no effect for them. Examples of respondents in this group are location-based services, for which source does not matter as long as the necessary performance is available, and the military, which does not rely on foreign PNT systems without memoranda of agreement (MOAs).

As for regulations limiting or prohibiting participation in a market, most of those who responded (responses were approximately equal between U.S. and non-U.S. users) did not think such regulations would be a hindrance. However, one respondent noted that given the complexity of the situation (the cost structure of the technology, the improvement of the technology, and the specific market segment), the answer was, "It depends." Some respondents suggested that the question of market participation called for a revisit of defense and national policies on satellite navigation/timing data. In particular, one respondent noted that until there is a clearly defined national policy on GPS/Galileo/GNSS, many U.S. developers would refrain from pursuing European business/participation because of technology-export licensing restrictions.

The majority of the respondents indicated that they were planning to support multiple systems, for one or more of the following reasons:

- Performance—increased availability, continuity of service, integrity of service, and ability to resolve ambiguities in service.
- Logical extension of business strategy—already supplying differential GLONASS service and will do so for Galileo if value added.
- Flexibility—reduced reliance on a single, military-controlled service.
- Growth—increased opportunities for services and software.

Some respondents were uncertain about their plans to support multiple systems of the GNSS, noting that whether they do so will depend on the cost benefit, which includes a host of pragmatic issues, such as the ability to define and use common standards. In this regard, the respondents offered a list of parameters they considered key to achieving effective interoperability in user equipment across the GNSS. The most frequently mentioned items were

- *Spectrum allocation and data modulation*—a shared spectrum, with orthogonal modulation schemes protected.
- *Time differential*—a common time frame.
- *Signal power*—for indoor use and when optimal antenna placement for reception is challenging. Those in markets 1, 2, 4, 6, 8, and 9 identified signal power as important, and several respondents believe that increased signal power will facilitate market expansion.

Finally, we asked the respondents to consider the significance of nonmilitary control of satellite-based PNT. This issue was not part of the "impact of competitive factors on demand" framework; it was added because our presurvey interviews and discussions suggested that it be explored with the respondents. The responses on this issue were equally divided: Some respondents thought it would have little to no effect; others thought it would have a positive effect. Those who saw the impact as positive expressed a common sentiment that nonmilitary control would help international business grow and a shared belief that improved reliability would result. According to one respondent: "It would improve our ability to sell into markets and cultures [that] are suspicious of relying on technology controlled by the U.S. military." Another respondent, however, suggested that there would be an adverse impact because "the volatility of market pricing for products/services provided by the civilian sector will limit the general market opportunity." It is worth noting here that, just as was true of finding new and novel applications for PNT data, it is thought that freedom from single control authority over PNT data will open new markets/products/services that otherwise would not emerge (i.e., as a result of performance improvements).

From these responses, we conclude that there is no one answer to the possibility of regulation, be it for mandating the use of Galileo or for restricting business opportunities. Mandating the use of Galileo—to the exclusion of GPS—seems to be a remote possibility to many of the experts to whom we talked, and none of the survey respondents expected to exclusively support one system or the other.

The respondents' business markets shaped their perspectives on restrictive practices. The markets most likely to be affected, according to our respondents, are the markets that are becoming more and more of a commodity product (e.g., receiver/chip manufacturing). And there is less motivation to enter into only these markets—for example, providing just the receiver may be a decreasing market option, according to Frost & Sullivan (1998, p. 1-7):

> The ability to introduce new products for more integrated use as well as offer complete navigation solutions has become an increasingly important factor for many companies as end-users seek single supplier options. This has led to an increase in acquisition and merger activity amongst companies who serve the general marine markets in particular.

We can also conclude that industry will try to use as much of the GNSS as it can make a business case for, and that the business case will be improved if there is cooperation about spectrum, reference systems (time), and power levels. Lastly, two factors were suggested as drivers of market expansion: (1) greater signal power, to support new and novel applications, and (2) an alternative control authority, to mitigate concerns in some regions over U.S. military control.

CHAPTER FIVE

Competitive Factors and Their Economic Implications

The relationship between GPS and Galileo is not one of either cooperation or competition, but, rather, of degrees, or levels, of cooperation and competition. When Galileo emerges, the environment for the two systems will consist of some level of cooperation and some level of competition, the extent and significance of which will have been determined by the resolution of many factors. Our study began by asking: How significant are these selected factors? What influence do they have on market demand (i.e., consumer surplus)?

Chapters Two through Four provide the basis for the following judgments, which use the framework of consumer surplus and competitive factors discussed in Chapter One.

How Significant Are the Competitive Factors?

Table 5.1 summarizes the current status of each of the six competitive factors and our assessment of each factor's significance for demand and cost. In addition, each assessment is coded to indicate how likely it would be to change if significantly greater numbers of survey respondents had completed the survey. Fuller explanations of our assessments are provided in the following paragraphs.

Table 5.1
Significance of Selected Competitive Factors

Factor	Status	Impact on Demand[a]	Impact on Cost[a]
(1) Interoperability/ Compatibility (Timing)	TWG working toward resolution—both systems provide data for consistency when using a combined constellation	None to insignificant (Δ)	None (Δ)
(2) Interoperability/ Compatibility (Geodesy)	Mitigation readily available—provides consistency when using a combined constellation	None to insignificant (Δ)	None (Δ)
(3) Interoperability/ Compatibility (Spectrum Sharing)	Galileo OS signals overlap GPS civilian signals at L1 and L5 (E5A) but not L2	Significant for L1 market; none for L2 market; insignificant for L5 market[b] ($\Delta\Delta$)	None for L1 market; potentially significant for L2 market;[c] insignificant for L5 market ($\Delta\Delta$)
(4) Strategies (Mandating Use)	Uncertain	Significant (D)	All ranges possible (Δ)
(5) Strategies (Regulating Industrial Participation)	Uncertain	None[d] (Δ)	Insignificant to significant ($\Delta\Delta$)
(6) Galileo Development and GPS Modernization	Uncertain	None for pre-Galileo IOC; significant for after Galileo IOC ($\Delta\Delta$)	Insignificant to none[e] (Δ)

[a] The notations in parentheses in this column represent the likelihood that our assessment of impact would change if significantly more respondents had completed the survey. Δ indicates little to no likelihood of change, $\Delta\Delta$ indicates some likelihood of change, and $\Delta\Delta\Delta$ indicates a great likelihood of change.

[b] These assessments are relative to L1. The smaller market for L5 (e.g., aviation) and long delay to use this frequency (GPS III is required for IOC) make the L5 impact difficult to assess.

[c] Several experts we talked to suggested that if only one system uses L2, the L2 user base may be small (manufacturers will seek cost-effective configurations), and the cost associated with L2 use will not benefit from economies of scale. Without a large L2 user base, the support needed to defend the spectrum from encroachment by other devices may be lacking, which may cause the cost of these receivers to increase to make them operable in what will become a noisy environment.

[d] Our assessment is that a user's decision to buy or not will be influenced not by what company is providing the data/services/products, but by what data/services/products are available.

[e] This may be a case in which a new environment stimulates many more providers to enter the market, possibly reducing user costs—similar to what occurred in telephony markets.

Factor (1) Interoperability/Compatibility (Timing)

GPS and Galileo use independent timing references. GPS references the Coordinated Universal Time (U.S. Naval Observatory) (UTC[USNO]), and Galileo will reference the Coordinated Universal Time (Bureau International des Poids et Mesures) (UTC[BIPM]). There are significant differences between the two for high-precision users, but the impact of these differences is being mitigated by actions of the U.S./EC GPS-Galileo Technical Working Group (TWG), which is attempting to develop a specification for GPS and Galileo so that one or both of the systems will broadcast offsets to the other.[1]

Industry input suggests that timing differences and the way in which they are resolved will have an insignificant effect on equipment cost.

Factor (2) Interoperability/Compatibility (Geodesy)

GPS and Galileo use independent geodesy references. GPS uses the World Geodetic System—1984 (WGS-84) coordinate system, and Galileo will use the International Terrestrial Reference Frame (ITRF). These systems currently agree to the centimeter range, and a mapping between the two can readily be accomplished in the receiver. Existing analysis indicates that the geodetic reference is not an issue (Hein, 2002). Industry input suggests that geodetic differences and the way in which they are resolved will have an insignificant effect on equipment cost.

Factor (3) Interoperability/Compatibility (Spectrum Sharing)

The ability to support different frequencies increases receiver cost and complexity, because extra or more-complex antennas, filters, and associated RF components are needed. User needs for multiple frequencies vary by market/application. Scientific and/or high-precision users are more likely to want to take advantage of multiple frequencies. Consumer items benefit from the additional satellites-in-view, so remaining with the L1 frequency would allow immediate use of

---

[1] The GPS/Galileo TWG, which first met in October 2002, is exploring technical issues that have implications for the potential compatibility and interoperability of GPS and Galileo.

Galileo, in addition to GPS, without having to change receiver electronics.

Factor (4) Strategies (Mandating Use)

The Galileo revenue model is based on two sources of revenue, both royalties: royalties from chip manufacturers and royalties from service providers employing Galileo in their service. If Galileo is to be successful at generating meaningful revenues, some believe that a market for Galileo-based products and services will have to be created via legislation, particularly to take advantage of high-volume markets in which opportunities for royalties and service revenues are greatest.[2] In broad terms, the economic impact will be artificially inflated demand with cost effects that may vary by market/application.

Factor (5) Strategies (Regulating Industrial Participation)

One of the stated motivations for pursuing Galileo is to develop European technical capabilities. This objective has raised concerns that opportunities for manufacturers and service providers will be restricted to European companies. This restriction will be challenged by companies seeking to expand their customer base by making their products compatible with all components of the GNSS (Frost & Sullivan, 2000), the emergence of complete navigation solutions, and the resultant increased merger and acquisition activity in some markets (Frost & Sullivan, 1998). Domain experts questioned whether it will be possible to limit participation for the OS, but they believe an attempt may be made for the SoL and PRS.[3]

---

[2] Our study did not assess whether enforceable intellectual-property rights, which are implicit in Galileo's revenue model, can be established. It also did not determine what revenue margins are feasible, given that GPS chipsets are becoming a commodity and alone are not very expensive.

[3] We did not assess whether these limitations could be effectively enforced. Our assumption is that enforcement would be accomplished through legal measures and technology (i.e., intellectual property rights and encryption) and by region.

Factor (6) Galileo Development and GPS Modernization

GPS modernization includes a series of acquisitions to provide IOC (defined as 18 satellites) and FOC (24 satellites) for the new civil signals, L2C and L5. According to the program schedules (see Appendix C), L2C reaches IOC after 8 IIR-Ms in 2QFY2007 and 10 IIFs in 4QFY2010 (note that the Enterprise chart, Figure C.2, indicates IOC in early 2009). L2C FOC occurs in 3QFY2012. L5 reaches IOC sometime after FY2012 because there are only 16 IIF launches by 4QFY2012 (note that the Enterprise chart, Figure C.2, indicates IOC in late 2011). FOC for L5 remains uncertain, because the schedule for GPS III and the remaining L5-capable satellites is not firm.[4] Galileo development (see schedule in Appendix C) will achieve IOC in 2008.

The effect of modernization or development will come down to this: What will the modernization or development accomplish and when? In the near future, Galileo may have an advantage in several areas:

1. *Increasing the number of satellites-in-view.* If receivers can use each additional Galileo satellite with little or no modification, the number of satellites-in-view and the corresponding availability will begin to improve with the very first Galileo satellite. Since GPS is sustaining an existing constellation—not adding on to that constellation—it does not have the same advantage.[5] If many

---

[4] As stated earlier, we did not directly assess the feasibility of this schedule. However, the existing launch policy, availability of launch vehicles, and overall program status (e.g., GPS III and the other segment acquisitions) raise concerns about meeting these dates. The current launch strategy is "launch on sustainment (or replenishment)" as opposed to "capability." But the satellites are lasting longer, an average of 10.8 years instead of their initial design of 7.8 years, thus pushing newer capabilities farther to the right. Without a change to the national policy, more-capable satellites are not going to be launched while a constellation of good satellites "able to support the user base" is in place (see http://spaceflightnow.com/delta/d295/). In addition to the space-segment changes, modifications to the new control station and full production of modernized user equipment will be required.

[5] Strictly speaking, replenishment may leave some satellites with residual capability, but we did not consider that here.

more applications emerge as a result of the performance improvements—more satellites-in-view—then demand may increase.

2. *Eliminating the single point of control.* Concern about the United States—more specifically, the U.S. military—and its sole control of satellite PNT will be alleviated with a Galileo constellation of sufficient size, once Galileo reaches IOC. If manufacturers, providers, and users find a non-U.S.-only environment appealing, demand may increase.

3. *More capability.* Galileo will provide new services that may open up new markets, and it may offer capabilities that GPS will not be able to provide (near term), such as signals with more power, higher-data-rate messages, and modern signal designs. These advantages require that Galileo reach IOC. GPS improved signal design begins in IIR-M. The next opportunity for GPS to build a new civil signal with higher data rates is GPS III. If providers and users value these capabilities, demand may increase.

What Influence Do the Competitive Factors Have on the Economic Benefit?

The domain experts and survey respondents in our study had different expectations for the future market growth of GNSS-related products and services. Some saw the market as already relatively mature; others thought that novel applications, increased performance (as a result of more satellites and increased power), or increased user confidence (notably in international markets) would increase market demand.

It is unlikely that the demand for location and time information will decrease. The ever-increasing number of applications that use location and time information, coupled with the miniaturization and reduced cost of devices and sensors, will sustain or increase the demand for this information. Multiple options for acquiring location and time information may also sustain or increase the demand. For instance, space-based alternatives (primarily GPS, currently) have supported significant growth revenues in such markets as car navigation, consumer electronics, tracking/machine control, and survey/mapping. Some markets may be able to take advantage of emerging means, such as network-centric positioning or Global System for Mobile communications (GSM)/Universal Mobile Telecommunication System (UMTS) ranging methods.

Galileo's entrance into the space-based PNT market may provide incentives for new applications because it will offer an option to the current military-controlled systems and will increase the constellation size. Constellation geometry is important for accuracy. Equally important is the ability to determine whether the position solution given by GPS is acceptable for use (integrity), an ability facilitated by a larger constellation size.

An increase in consumer surplus results from an increase in demand while the cost remains constant or decreases. Table 5.2 summarizes the *potential* impact on consumer surplus of each competitive factor independently, based on the analysis of each factor described in Table 5.1. A factor that produces an increase in demand while cost remains constant or decreases will produce an increase in consumer surplus. A factor that produces an increase in price while demand remains constant[6] will produce a decrease in consumer surplus.

As can be seen in Table 5.2, we conclude that two of these factors—L2 spectrum sharing and regulating industrial participation—have the potential to cause a decrease in consumer surplus.

With regard to the first factor, a lack of commonality in the L2 frequency may mean that civilian users' costs to use this frequency will go up to cover extra costs for the measures that may be required to protect the L2 frequency from spectrum encroachment and/or extra costs stemming from diminished economies of scale.

And as for the second factor, the lack of an open and competitive market rules out reduced costs through competition. When the entry barriers for a market are high (such as restricting companies

---

[6] No option considers the case for which demand decreases. We did find evidence in selected markets, such as telecommunications, that alternative options are being explored—a limited approach at present.

Table 5.2
Impact of Competitive Factors on Consumer Surplus

Factor	Consumer Surplus		
	Increased	Decreased	No Change
(1) Interoperability/ Compatibility (Timing)			Nonissue
(2) Interoperability/ Compatibility (Geodesy)			Nonissue
(3) Interoperability/ Compatibility (L1 Spectrum Sharing)	After Galileo IOC, mainly due to availability and civilian control		
(3) Interoperability/ Compatibility (L2 Spectrum Sharing)		Lack of commonality	
(3) Interoperability/ Compatibility (L5 Spectrum Sharing)	After Galileo IOC, mainly due to availability and civilian control		
(4) Strategies (Mandating Use)	In regions of EU control or influence		Outside regions of EU control or influence
(5) Strategies (Regulating Industrial Participation)		Mainly due to lack of open and competitive market for provision of goods and services	
(6) Galileo Development and GPS Modernization	After Galileo IOC, mainly due to availability and civilian control		

from participating), fewer companies will participate, and the ones that do will have little pressure on them to be competitive (in terms of price or service). This condition is likely to have a negative effect on price and demand, resulting in a smaller consumer surplus.

In both cases, the severity and duration of the effects are uncertain. Therefore, the United States should continue to monitor these factors and prepare strategies to mitigate their effects. Greater cooperation and joint planning for future improvements with the EU may lessen the L2 spectrum issue. Policies that allow U.S. firms to actively participate, including less-restrictive export licensing, may encourage U.S. firms to participate and allow them to compete globally.

CHAPTER SIX

The Economic Impact: Market Response to Galileo and What the United States Should Do

Right now, GPS is the one "full-service" GNSS system for civilian use. It has a satellite constellation that broadcasts civilian signals on one frequency, L1, and it can be augmented by GLONASS, space-based augmentation systems, and land-based augmentation systems. The GPS has two important sets of stakeholders. The first set consists of the system's users, a group that has two components: consumers who derive benefits from consumer products (goods or services) that use GPS signals, and producers (not including providers of GPS-related products) that use GPS information to produce their goods and services more efficiently. The second set of stakeholders consists of providers of GPS-related products ranging from receivers, to sophisticated goods that incorporate receivers (such as car or aviation map/navigation equipment) and associated software and data upgrades. The U.S. military is, of course, an important stakeholder and properly fits in both sets; indeed, the system was originally primarily designed for and produced by the military.

The Europeans have now committed to producing a second full-service system, Galileo. It will be a separate constellation with similar functions but different capabilities, and it will broadcast on frequencies that are both the same as and different from those of GPS.

In this chapter, we broadly consider what Galileo implies in terms of market demand and the coexistence of GPS and Galileo. We begin with a simple model to represent the market demand impact. Then we present three scenarios describing alternative future worlds, as they may be once Galileo exists. Finally, we describe several options for the United States in response to these future worlds.

Market Size

We begin with users of the system and consider how they might behave in both the presence and the absence of Galileo. Assuming a year in the future in which Galileo has, in accordance with current plans, reached FOC, we compare that world with a world in which Galileo has not been produced. Table 6.1 shows how we define the size of the market. We postulate that if Galileo were not produced, the size of the GPS market (i.e., amount of GNSS-related goods and services produced) in the future year would be X, where X is defined as an index (constant-dollar amount) of the level of production of GPS-related goods and services.

   In the combined GPS-Galileo world, there would be three distinct markets: GPS-only users, Galileo-only users, and combined users. The size of each of these markets is represented by the variables Y, L, and C, respectively, which means that the total size of the GPS-related market is $Y + C$, and the total size of the Galileo-related market is $L + C$. Thus, the existence of the Galileo system is likely to change the size of the GNSS market.

Table 6.1
Size of GNSS Market in Two Cases

GNSS Market	Case	
	Galileo Not Available	Galileo Available
GPS-only users	X	Y
Galileo-only users	—	L
Combined users	—	C
GPS market	X	$Y + C$
Galileo market	—	$L + C$
Total GNSS user volume	X	$Y + C + L$

Implications of Various Futures

In a world in which Galileo is available, how would users decide which option to choose? Each option is going to have an associated cost of receiving the signal and an associated signal quality. We characterize three general kinds of future worlds qualitatively here. (Appendix D provides quantitative characterizations; however, since the data useful for this purpose are not all available, these characterizations are necessarily preliminary.) These worlds are based on the assumption that those who would have used GPS in a GPS-only world will now, with Galileo available, have three choices:

1. Continue to use GPS only.
2. Use a Galileo-only system.
3. Use a combined GPS-Galileo system.

We discuss each of these choices in turn.

Continue to Use GPS Only

For GPS, the cost of receiving the signal is the cost of the receiver. Since users have chosen to use GPS in the absence of Galileo, the combination of receiving cost and signal quality must be of value to them (i.e., superior to not using a signal at all and thus not incurring the receiver cost). Choosing to continue using GPS only (for the civil user) would imply that the user does not value the receiving cost and signal quality of either the Galileo system alone or the combined GPS-Galileo system over that of the GPS system alone.

Use a Galileo-Only System

Using Galileo alone also has an associated receiving cost and signal quality. The assumption is that the Galileo receiver cost will be higher than that for GPS, since Galileo management plans to levy charges on some signal users and, potentially, on equipment manufacturers. Therefore, the resource cost of producing the receivers for either system's signal will be the same. If the resource cost of producing the

Galileo receiver were sufficiently lower, however, the above statement about relative costs could be reversed.[1]

We propose two reasons for why a user would decide to use a Galileo-only system. First, the user would *voluntarily* choose this system based on the judgment that its receiving cost and performance package are superior to those of both the GPS-only system and the combined system. In this case, the United States has three choices:

1. It can accept this outcome and let GPS revert to a military-only system.
2. It can engage in technical/market research to determine why users prefer Galileo and then decide whether GPS upgrades to attract users back to GPS (or at least to a combined system) make economic sense. Appendix D illustrates how such a decision might be considered.
3. It could mandate GPS use in order to support the U.S. GPS-related industrial base.

We recommend against the last policy, both because it would lower the overall economic welfare in the United States and because it would likely trigger retaliation by Europe.

The second reason that a user would choose a Galileo-only system is that European authorities mandate it. In this case, harm might be done to the U.S. industrial base for GPS-related-products if the U.S. GPS industrial base were prohibited from participating in the Galileo market. Harm might come to other industries, too, such as transportation and shipping, if they were forced to modify their equipment to support Galileo in one region of the world and GPS in another.

---

[1] Throughout our discussion of receiver costs, we of course have to recognize that receiver costs are likely to be a decreasing function of both cumulative quantity produced and annual quantity produced. Thus, when we speak of receiver costs being "higher" or "lower," we really mean "on a higher cost function" or "on a lower cost function." The actual cost of receivers that occurs will then depend both on the cost function and on the size of the resulting market, as measured by both cumulative and annual production.

Here, again, we recommend that technical/market research be conducted so that both the United States and the EU can better understand what kind of outcomes to expect. We also, again, recommend that GPS and Galileo technology be coordinated (provided non-EU manufacturers are not prohibited from participating) so that combined-system receivers do not become excessively costly and so that the signals from both sets of satellites can be used together to provide users with the highest quality of information. Such coordination should lead to a future in which the use of combined signals is the voluntarily chosen path of the market, and it will not only maximize the economic welfare of all users, but also lead to vigorous industrial bases of both GPS-related and Galileo-related production. Obviously, the United States has a variety of retaliatory policies that it could use as well. We recommend that the United States use those policies only as a last resort but use the *possibility* of those policies to induce a negotiated end to this kind of protectionist policy. Appendix D indicates which of these policies could be used and how they might change the payoff to European authorities so as to induce them to return to a free market.

Use a Combined GPS-Galileo System

We expect that a receiver capable of receiving both signals will cost no less than a receiver capable of receiving only one signal, and that its cost will increase with the number of frequency bands received. As discussed in Chapter Five, a current cost driver for receivers is the number of frequencies they must support, since extra or more-complex antennas, filters, and associated RF components are required for the additional frequencies.

However, even if a receiver for a combined GPS-Galileo system costs more than a receiver for a single system, users may choose a combined GPS-Galileo system because of its benefits, which fall broadly into two categories: performance improvements and effectiveness improvements.

Availability, position accuracy, and integrity all benefit from a combined (i.e., larger) constellation. Also, having more satellites-in-view provides improved performance, which may allow users to do without ancillary equipment and services, such as space-based augmentation or high-stability oscillators for timing applications. This could translate into cost savings that make the increased cost of a combined-system receiver very acceptable.

If we judge that the benefits listed above outweigh any receiver cost, we might expect that additional users (including, of course, additional uses by existing users) will also buy into the combined-system market, because of the system's increased technical performance and increased robustness, and the greater user assurance that it offers compared to a single system. Using GPS is now somewhat of a case of putting all one's eggs in one basket, whereas using a combined system adds insurance against this case and should make a multisystem GNSS a more attractive product.

This third, combined-system outcome is very attractive for the United States. It offers GNSS users more economic benefits than they now get, and it offers providers of GPS-related products an expanded market. U.S. policy should, all other things being equal, attempt to realize such an outcome. To this end, the important policy aspects are those that focus on coordinating GPS and Galileo technology so that combined-system receivers are not excessively costly and signals from both sets of satellites can be used together to produce information of the highest quality for users.

The Most Likely World

Our research suggests that the last world—the one in which a combined system is used—is the most likely and that the products and services offered to the user will probably be a combination of GPS and Galileo products. The assumption underlying the current notion of competing is that the user will choose one system over the other. However, PNT equipment suppliers and service providers have indicated that in order to remain competitive, they are considering incorporating all data sources (including augmentations beyond GPS and Galileo) when it is cost-effective to do so and provided they have access to the necessary technical information.

CHAPTER SEVEN

Conclusions and Recommendations

For this study, we examined factors of the competitive environment in order to understand how the Galileo system might affect the current set of GPS stakeholders. Our ultimate aim was to recommend U.S. policies that will produce outcomes favorable to the stakeholders in a world where GPS and Galileo coexist. We examined a portion of a multifaceted problem that has technical, geopolitical, regulatory, national security, and economic dimensions. We intentionally focused on civilian and economic matters rather than on national security matters, which, for the near future, will significantly influence any U.S. actions with respect to GPS and Galileo.

In this chapter, we provide our conclusions regarding the economic impact of Galileo and the implications thereof for the United States. We also provide recommendations for the United States that are consistent with increasing the economic benefit and that can leverage the existence of Galileo to offer enhanced capability for the user.

What Is the Economic Impact of Galileo from the U.S. Perspective?

The demand for location and time information is unlikely to decrease. It will stay the same or increase as a result of the ever-increasing number of applications that use this information, coupled with the miniaturization and reduced costs of devices and sensors.

Having different options for acquiring location and time information may also sustain or increase the demand, and Galileo is one such option. Thus, the existence of the Galileo system as planned may change the size of the market by accelerating the demand. We have concluded that of the factors we examined, only two—L2 spectrum sharing and regulating industrial participation—may produce negative economic effects and that there are ways to mitigate these effects.

Even by matching GPS capabilities, Galileo may produce increased interest in satellite-based PNT. For the largest and fastest-growing market segment (consumer applications), the required functionality and performance of the two systems are fairly comparable. Studies that examined combined Galileo-GPS system configurations (*Galileo Services and Architecture*, 2002) showed them providing only modest increases in accuracy. However, users see the existence of the two systems as beneficial, largely because the combination offers additional satellites-in-view, which will provide better availability and a more robust architecture to support the integrity function. Users also see as beneficial the fact that Galileo as planned will be controlled by an alternate, nonmilitary organization.

Recommendation: The United States should remain indifferent to Galileo, from an economic standpoint, as long as the EU does not apply restrictive policies/regulations. U.S. responses to such restrictions could include retaliatory practices (e.g., mandating GPS), providing a superior civilian service based on market research, and increasing cooperation with Galileo. We do not recommend the first action; we view the second and third actions as more likely to result in an increased net economic benefit.

What Conditions Will Have Favorable Economic Benefits?

A more cooperative approach, one that allows users to seamlessly use both PNT systems, is more likely to result in conditions favorable to users and to be supportive of innovation as a result of potential performance improvements and improved user confidence. However, these improvements are tied to the nature of cooperation and competition that will exist between the systems.

The challenges to greater cooperation and increased coordination between GPS and Galileo are nontrivial, largely because of the lack of appreciation for the differences between the U.S. and EU objectives that, until recently, has been an impediment to effective interaction. The current scope of cooperation—which is at the user level, where success is considered to be achieved as long as the systems do no harm to each other—misses opportunities for greater benefits. By focusing on mitigating differences, the providers are delaying the opportunity that cooperation offers for efficiency in the provision of PNT data/services. A greater level of cooperation in providing PNT data/services may have other benefits as well—for example, it may discourage others from proliferating GNSS standards, and it may establish a stronger constituency to protect the GNSS spectrum.[1]

Greater benefits for users are possible if GPS and Galileo cooperate so that the two systems are combined (appear as one). The primary benefit stems from having more satellites-in-view and shared frequencies for simpler receiver design. Combining the two systems facilitates increased performance in accuracy (having more satellites helps the geometry problem), integrity (having more satellites helps determine whether the system is providing reliable information), and continuity (having more satellites helps when there is a sudden loss). And when the two systems truly appear as one, with shared frequencies, simpler and less costly receivers can be used.

There may be interesting opportunities for the United States to employ both GPS and Galileo signals to improve PNT performance, which could enable more cost-effective and value-added use of augmentation systems. In our research, augmentation systems providers indicated that Galileo presented a real threat to their existence. They noted that they might have to offer some other value-added service if Galileo were to cause the functionality and performance level of their

---

[1] "The U.S. has the leadership in most of the military space systems and there is no denying that it sets the standards. But what happens when the standard is not shared? Another standard will emerge to further reinforce walls and fortresses" (Sabathier and Sapolsky, 2002).

augmentations to become obsolete. This is a consideration for all augmentation systems, including those managed by the IGEB, such as WAAS.

The key barriers for greater U.S.-EU cooperation and coordination in providing satellite PNT are primarily political, not technical. The United States and the EU have similar but competing economic objectives for PNT data/services. The United States wants to encourage private sector investment in and use of U.S. GPS technologies and services, to advance U.S. scientific and technical capabilities, and to promote commercial market growth and trade. The EU wants to develop European technical capabilities, to create new jobs in Europe, and to build a market position for European firms. Furthermore, the EU and the United States have distinct perspectives on national security and strategic objectives that lead them to view the operation and use of these highly valuable, dual-purpose assets differently from each other.

Recommendation: The United States should directly address the political impediments to greater cooperation in order to explore the range of options for bringing about greater opportunities in providing PNT data/services. It is important for the United States to improve the perception that GPS is a trustworthy and reliable resource for the global community, to leverage opportunities (such as Galileo) to modernize GPS and offer enhanced augmentation services, and, potentially, to maximize GPS's use for future coalition operations. Working with the EU as a cooperative partner in the provision of PNT data/services may help attain these goals.

What Are the Implications for the United States?

A combined environment, where users are able to easily use both systems, implies more economic benefits for users of GNSS than they now get and an expanded market for providers of GPS-related products. According to our research, this is the most likely of the three future worlds, but its realization involves challenges.

At the heart of the understated challenge for GPS is this fact: As a dual-use system, GPS must serve two primary purposes—economic growth and national security. Actions taken to be competitive for economic growth are not necessarily consistent with actions for national security. Economic growth requires open standards and ongoing improvement to encourage broad adoption and recurring purchases; national security requires confidentiality and protection to ensure authorized use only, and improvements are dictated by operational military objectives. Economic growth is enhanced by industrial partners perceiving PNT as an open and reliable resource. National security is enhanced by creating cooperative arrangements with coalitions and allies to protect sensitive information.

The ability of GPS to effectively serve the objectives of national security *and* economic growth is becoming more and more of a challenge. In light of Galileo, U.S. actions taken to promote civilian use and economic growth may be inconsistent with actions necessary to retain a national security advantage. We note that a future world in which civilian users voluntarily migrate to or add on Galileo services may present the United States with an opportunity to avert some future expenditures for GPS civilian requirements. We also note that there is a point at which GPS providers will have to take a position on what level of performance and functionality to commit to for the civilian user.

Recommendation: The United States should reevaluate the *implications* of GPS's dual-asset nature. Clearly, GPS is and will remain a dual-use system, but a potential opportunity exists to improve the civilian service in ways the United States can do only if it shares the burden. Should the United States seek to formally share the responsibility of satisfying civilian user needs with the EU? Included in this decision is another one: What level of commitment will GPS providers offer to the civilian user base above and beyond what is currently offered? Both the GPS and the planned Galileo system are trying to provide a level of robustness and service that is difficult to meet individually but may be more easily achieved jointly. A combined system may allow both the United States and the EU to provide high performance and robustness without maintaining the current 24+ satellite constellation at all times. This possible scenario—combined, co-operating GPS and Galileo systems—should be examined in earnest but raises many additional questions that require further analysis and evaluation, such as: How much U.S. independence is needed and how much interdependence is tolerable, particularly for national security concerns? What metrics are available for assessing how well these changes would meet U.S. national security objectives, missions, and concerns? What assurances would be required of the EU to demonstrate its commitment as a reliable partner capable of developing, deploying, and sustaining the Galileo constellation over time? What would be the impact on the many and diverse augmentations that have emerged to satisfy the growing civilian need?

Industry Participation

Table A.1
Survey Respondents, Their Market Segments, and Their Products/Services

Survey Respondent	Market Segment	Product/Service
Nokia Mobile Phones (Finland)	2	Manufactures mobile phones with built-in GPS receivers to meet mandatory E911 requirements in the United States
Leica-Geosystems (SW)	3	Manufactures GNSS receivers for high-precision (cm positions) survey market
Time and Frequency Solutions Limited (UK)	9	Uses commercial, C/A code GPS receivers within precision instruments to synchronize timing signals and to discipline oscillators for high-stability frequency reference applications
Symmetricom (U.S.)	9	Designs, builds, and sells precise time and frequency equipment for applications such as telecom network synchronization and test and measurement, most of which use GPS as primary reference
Precision Timing Solutions (U.S.)	9	Uses GNSS signal to precisely discipline internal clock of a GNSS receiver for timekeeping, timing, location, positioning, and network-synchronization applications
University Technologies International (Canada)	1, 2	Conducts R&D, software development, and equipment testing for various civilian and military markets
OmniSTAR Europe BV (Netherlands)		
NavCom Tech (U.S.)	6, 4, 3	Supplies high-accuracy GPS receiver and products for agricultural, offshore, survey, GIS, and machine control applications; and supplies GPS satellite correction service capable of real-time decimeter accuracy worldwide

Table A.1—continued

Survey Respondent	Market Segment	Product/Service
Rockwell Collins (U.S.)	8	Produces satellite navigation equipment for global users in military and government agencies
Advanced Research Corporation (U.S.)	8	Assesses GPS constellation performance (past and future)
Honeywell Aerospace Electronic Systems (U.S.)	5	Supplies aviation electronics and receivers for Air Transport & Bizjet aircraft; manufactures avionics systems (i.e., IRS, FMS, EGPWS, GPS navigators, recorders, and primary flight displays) that use GPS position, velocity, altitude, time, and time marks from GPS receiver(s)
Fugro (Norway)	7, 3, 4	Provides high-performance services, such as positioning for station keeping or vessel docking, co-tidal height measurements, and machine guidance
FieldWorker Products Ltd	3	Supplies handheld units to store location information for consolidation by such users as geologists, national parks, local governments, farmers, and many commercial enterprises and to navigate on maps or to known points
Science Applications International Corp (SAIC)	8	Uses GPS services as an enabling technology in applications for military, civil commercial, and governmental customers; and provides acquisition and advanced technology development and test and evaluation support to both DoD and DOT
GMV Sistemas PTM S.A. (Spain)	4	Provides a positioning service included in applications for fleet management solutions
SPIRIT Corp. (Russia)	6	Manufactures GPS/GLONASS software receivers
Telcontar (U.S.)		Provides location-based services, including mapping, navigation, real-time traffic, fleet monitoring and homeland security, vehicle and asset management, and concierge services
Linkspoint (U.S.)		Designs and manufactures GPS receivers, integrates GPS into enterprise applications, and develops software products that use GPS
Meinberg Funkuhren (Germany)	9	Synchronizes timing and frequency

Table A.2
Companies Interviewed Directly and Their Products/Services

Company	Product/Service
Trimble	Provides advanced GPS components and augments GPS with other positioning technologies and wireless communications for growth in applications including surveying, automobile navigation, machine guidance, asset tracking, wireless platforms, and telecommunications infrastructure
NAVSYS	Provides specialized GPS products and services to include GPS hardware design, systems engineering, systems analysis, and software design for both governmental and commercial customers
Stansell Consulting	Provides consulting services to GPS Joint Program Office

Study Survey

- Car Navigation
- Consumer/Recreational
- Survey/Mapping/GIS
- Tracking/Machine Control
- Aviation
- Original Equipment Manufacturing
- Marine
- Military and Public Safety
- Timing

1. Using the market segmentation above, please select the group represented by your group/organization. If there is more than one, please select the primary group and use that group when answering the remaining questions. User group/organization
_____.

2. Please briefly describe how your group/organization uses satellite navigation data.

Please use the following definitions for answering the series of questions below:

Position Accuracy is a statistical value of the error between the true position and estimated position. Current specifications state that at the 95% confidence level the position accuracy shall be no greater than 4.0 meters horizontal and 7.6 meters vertical.

Availability is the percentage of time that the position accuracy meets the specified accuracy performance level.

Continuity Gap is the maximum continuous length of time that the specified position accuracy is not met without advance notification.

Integrity is the ability to determine whether the system is providing reliable navigation information. It is measured as the rate at which the system will not provide the user with hazardously misleading information (e.g., $X * 10^{-Y}/$ second).

Time-to-Alarm is the length of time required to provide notification at the user interface that the service is unavailable.

Timing Accuracy is a statistical value of the error between the true time (UTC) and the estimated time. Current specifications state that at the 95% confidence level the timing accuracy shall be no greater than 20 nsec for static user and 35 nsec for dynamic user.

Guarantee is the concept of ensuring services for applications in which a disruption of service would have significant safety-of-life or economic impacts.

3.   Please indicate the criticality of these parameters to your group/ organization.

	essential	somewhat important	not important
Position Accuracy			
Availability			
Continuity Gap			
Integrity			
Time-to-Alarm			
Timing Accuracy			
Guarantee			

4. For those parameters that are important to your group/organization, please define the performance/service you currently receive. If that performance/service is achieved via the assistance of an augmentation, please identify the augmentation service in the last column.

Performance Metric	Current Performance Realized	Augmentation
Position Accuracy	Vertical = Horizontal =	
Availability	Percentage of time =	
Continuity Gap	Length of time =	
Integrity	System reports to user hazardously misleading information at a rate of _____ /sec	
Time-to-Alarm	Length of time =	
Timing Accuracy	Static = Dynamic	
Guarantee	Yes_____ No_____	

5. Does your group/organization need improved performance/service in any listed parameters? If so please indicate what performance would be desired.

Performance Metric	Desired Performance
Position Accuracy	Vertical = Horizontal =
Availability	Percentage of time =
Continuity Gap	Length of time =
Integrity	System reports to user hazardously misleading information at a rate of _____ /sec
Time-to-Alarm	Length of time –
Timing	Static = Dynamic
Guarantee	Yes_____ No_____

6. Do you have a roadmap that will take you from the current performance level (question 4) to your desired performance level (question 5)? If yes,

   • Does this roadmap define the relationship between performance improvement and market growth?
   • Can you share this roadmap with us?

7. Please briefly describe how your organization would benefit from improved satellite navigation data as indicated above (in question 5).

8. If the higher performance levels (in question 5) were available today, how much larger (in percent terms) would you expect total sales in the market segment you are now in (see question 1) to be?

9. If these higher performance levels resulted in an increase in total market sales for your market segment (question 8), how much higher (in percent terms) would the prices of the products/services in your market segment have to be in order to reduce these total sales back to their current level?

10. In the future certain implementations of GNSS may be required for use in some market segments and selected regions. What impact, if any, would such a mandate have on your market segment? Do you anticipate that such actions might inhibit your participation in this market? If so, how?

11. The future GNSS will be composed of systems managed by the military and civilian sectors. What impact, if any, will increased civilian management of GNSS capabilities have in your market segment?

12. Would you use multiple GNSSs in your terminals or services? What would be the motivation?

13. (OEM only) In order to achieve good interoperability across multiple GNSSs, what are the key signal parameters (in priority order) that should be included in common user equipment standards? What is the approximate additional cost to your equipment (in percentage) if the multiple GNSSs do not standardize on these parameters?

14. Is there anything else important for your market but not mentioned above (e.g., signal power, etc.)?

GNSS Program Schedules

Figure C.1
Galileo

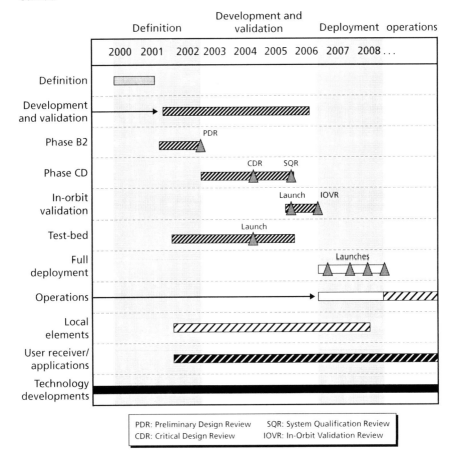

Figure C.2
GPS Enterprise Perspective Schedule

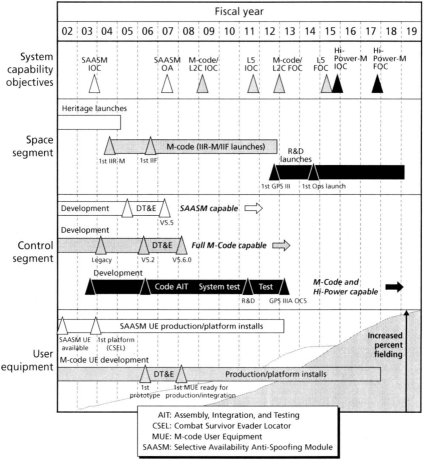

RAND *MG284-C.2*

Figure C.3
GPS Block IIR-M Schedule

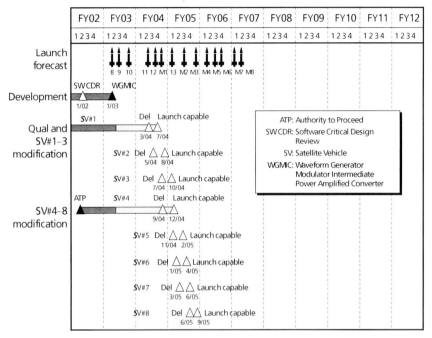

Figure C.4
GPS Block IIF Schedule

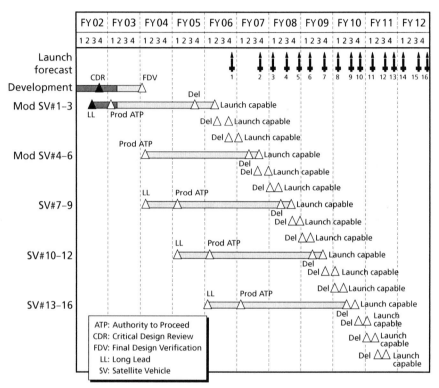

APPENDIX D

Analyzing the Economic Benefit

This appendix provides an economic framework to quantitatively analyze the economic benefit of GPS alone, Galileo alone, and GPS and Galileo combined. The concepts of economic benefit and consumer surplus are explained and an exemplar application of the framework to the GPS-Galileo question is provided. Although we mention in Chapter Seven that economic growth is one of the primary purposes of GPS, a careful analysis of growth should focus on the change in income (or gross domestic product, GDP) that occurs over time. GPS and GNSS can support this by increasing total factor productivity through network effects, etc. Further analysis of the growth implications has not yet been undertaken and would require a thorough case-oriented empirical analysis of the growth implications of the alternative forms of industrial organizations that might support GNSS.

In this analysis, we present a complementary approach that focuses on the static benefits and costs associated with these space-based systems.

Since both gross benefits and costs occur over time, we must consider the total future time stream of each, up to some future time horizon. To describe the economic impact of any system, the total gross benefits over time and the total costs could simply be added up. But a more economically meaningful measure is the present value of future benefits and costs.

The *present value* of any given future monetary value is the amount of money that would have to be invested today, at current interest rates, to generate that given amount at the given future date. Thus, the present value of all the future gross benefits of a system is the sum of the present values of each future year's gross benefits. Similarly, the present value of all the future costs of a system is the sum of the present value of each future year's costs. The *net present value of the system* is then defined as the present value of all the future gross benefits, less the present value of all the future costs.

The costs of any satellite PNT system are conceptually straight-forward to measure using a standard framework of R&D (both initial and continuing), procurement (both initial and replenishment), operating and support (O&S) costs, and disposal. Potential difficulties in determining these costs can arise with regard to highly sensitive components of the program or the problems inherent in allocating some kinds of R&D-related expenditures to any specific project. For example, general space-related R&D or certain overhead expenditures in R&D organizations may contribute to PNT systems, but they may not be allocated officially to any project. If such costs are accounted for differently by different organizations, comparison of their R&D costs may be biased.

For the GPS and GNSS markets, one would expect individual producers to incur setup costs that then permit the units to be produced and constant or, possibly, declining marginal costs. This suggests that the appropriate model for investigating costs and benefit is one that has a monopolistically competitive market structure—that is, a situation in which there are many differentiated sellers. Table 2.2, in Chapter Two, suggests that the PNT markets are car navigation, consumer/recreational, survey/mapping/GIS, etc., so the expectation is that the competition would be extensive and that intra-industry trade would emerge between the United States and the producers in a European consortium.

At this time, however, it is difficult to analyze the qualitative features of the market outcome using the monopolistically competitive model. We thus chose to provide some insights using traditional supply-and-demand analysis. One should recognize that the supply-and-demand model does not require that the units produced be identical, but that perfect substitution among the products within a submarket can occur. Furthermore each producer is viewed as having an eventually upward-sloping marginal cost curve.

Although we do not expect these assumptions to be literally satisfied, supply-and-demand remains a useful analytical construct when the actual market structure does not fully satisfy the underlying assumptions.[1]

With this qualification, we represent the annual market for the specified product by a supply-and-demand diagram, such as that in Figure D.1. That is, the quantity variables associated with the product will be annual levels of production and consumption. A different supply-and-demand diagram will exist for each future year. The diagrams will differ as a result of overall economic growth, changes in

Figure D.1
Market Supply and Demand

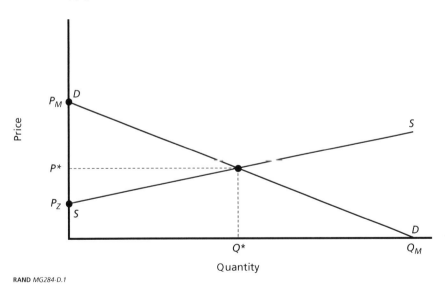

RAND *MG284-D.1*

---

[1] For example, one might choose to analyze the broad features of price changes in the automobile market using the supply-and-demand construct, as a first approximation, even though the underlying assumptions may not be satisfied.

technology, changes in tastes, or other relevant factors that change over time. U.S. government directives indicate that cost-benefit analysis should be conducted from the standpoint of U.S. citizens. Therefore, the indicated demand curve is for U.S. consumers. The supply curve is that facing the United States.[2]

Interpreting the Supply-and-Demand Diagram

To make this analysis more concrete, we discuss Figure D.1 as though it represents the annual market for handheld personal GPS receivers by U.S citizens, one part of the total S (space-based-PNT-enabled-products [SPEP]) market.[3] For clarity, we discuss the diagram as though it is for the year 2003 (although the same kind of discussion would apply for any future year). The horizontal axis represents the annual quantity of receivers produced and consumed (bought); the vertical axis represents the price of receivers.

---

[2] See Circular A-94, "Guidelines and Discount Rates for Benefit-Cost Analysis of Federal Programs" (www.whitehouse.gov/omb/circulars), which states (p. 5): "Analysis should focus on benefits and costs accruing to citizens of the United States in determining net present value." With respect to the supply curve facing the United States, this equals world supply less European demand: If Du represents U.S. demand, De represents European demand, Su represents U.S. supply and Se represents European supply, then the supply curve facing the United States is Su + Se − De. This holds because, in market equilibrium, Du + De = Su + Se.

[3] The disaggregation must be a true partition of the market—that is, it must include all SPEP but must not double-count. SPEP markets include all goods that incorporate a receiver for space-based PNT signals and all services that require reception and use of such signals for their production. Say that a given service sector that uses space-based PNT signals has a market of Y per year and that, each year, producers of that service purchase X worth of space-based PNT signal receivers for use in producing the service. $(X + Y)$ is obviously an overestimate of the SPEP market in this sector, since the X of goods purchased are then used to produce the Y of services. In this case, we must decide whether we are going to define the relevant market sector as the service itself, with an annual size of Y, or the particular receivers, with an annual size of X. If we choose the former definition, we must be sure to exclude the relevant receivers from all other market sector definitions. In general, market sectors must be very carefully defined to avoid this kind of potential mistake.

Line DD, the U.S demand curve, represents the amount of receivers that would be purchased by users per year as a function of receiver price. (The functionality of this line is thus that quantity, on the horizontal axis, as a function of price, on the vertical axis.) Point P_M shows the price that is so high that no receivers would be purchased. Point Q_M shows how many receivers would be "purchased" (i.e., accepted) by users if they were free.

Line SS is the supply curve facing the United States, which represents the number of receivers that would be available to U.S. consumers each year as a function of price. Point P_Z shows the price at which no receivers would be produced, and the upward slope of SS indicates that supply is higher at higher prices.

Price P^* and quantity Q^* are the market-equilibrium price and quantity, respectively. They are the price and the volume of consumption and production that will prevail in the market. At any price below P^*, the amount that individuals wish to purchase (from line DD) will be more than production (line SS), which will drive prices up. Similarly, at any price above P^*, the amount that individuals wish to purchase (from line DD) will be less than production (line SS), which will cause prices to fall.

Gross Economic Benefit and Consumer Surplus

We now use the supply-and-demand framework to identify the first kind of gross economic benefit that accrues to users of SPEP. Figure D.2 does this most easily. It shows a demand curve DD with a shape somewhat different from that in Figure D.1. It also shows a horizontal supply curve—that is, a supply relation in which any amount can be purchased at any given price, P^*, shown.

Figure D.2's demand curve DD explicitly shows the price at which each individual unit of annual production would be purchased. It is a step function in which each step has the width of one unit of production. Thus, P_1 is the price at which exactly one unit of the product would be purchased, P_2 is the price at which exactly two

Figure D.2
Derivation of Consumer Surplus

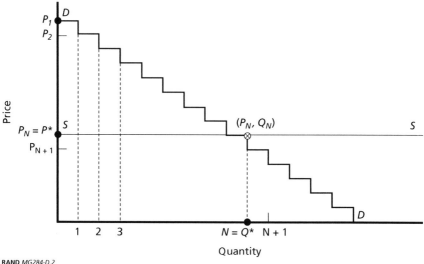

RAND *MG284-D.2*

units of the product would be purchased, and so on. We interpret P_1 as the value to the buyer of the first unit of production on the horizontal axis. It is clearly the most that the buyer would be willing to pay for the unit, since at a price above P_1 the unit would not be purchased. Similarly, the buyer must value one unit of the product at no less than P_1; otherwise, the buyer would not purchase it at that price. Therefore, P_1 is a measure of the value of the first unit to its buyer, which we also refer to as "willingness-to-pay." However, the buyer of that unit pays only P^* for it, since P^* is the market price, which means that the buyer of the first unit is acquiring a product worth P_1 to the buyer at a price of P^*. The difference between the value and the cost of this item, $(P_1 - P^*)$, is called *consumer surplus* and is interpreted as the economic benefit that the buyer receives as a result of participating in this market.

A similar argument can be made for the buyer of the second unit on the horizontal axis. The value of that unit to its buyer is P_2, since P_2 is the most the buyer would be willing to pay for it. Since this unit is also purchased for P^*, its buyer receives a consumer surplus of $(P_2 - P^*)$, which is similarly interpreted as the economic benefit that the buyer of the second unit receives as a result of participating in this market.

Similar arguments apply to the buyer of each unit, so that total consumer surplus, or benefits to buyers participating in the market, equals the sum of the values $(P_n - P^*)$, for $n = 1, \ldots Q^*$, total market production. Of course, $Q_N = Q^*$. Consumer surplus is then, equivalently, the area between the demand curve and the line representing market price.

We make two more observations about the definition of *consumer surplus*. First, the buyer of the Nth unit of production, the last one purchased on the horizontal axis, actually receives no economic benefit from participating in the market. That unit is purchased for exactly its value to its buyer.

The second observation about *consumer surplus* is as follows. Let us say that the market of Figure D.2 will disappear unless there is an annual subsidy of some amount. What is the most society should be willing to pay annually in such a subsidy to maintain this market? It is evidently the amount of consumer surplus: the amount that could be taken away from consumers and that would be the same as denying them the right to participate in the market. In the same way, *consumer surplus* is defined as the most that society should be willing to pay to bring such a market into being.

Figure D.3 is a reproduction of Figure D.1 but with the amount of consumer surplus indicated. For simplicity of presentation, most market diagrams show smooth demand curves (such as the one in Figure D.1) rather than stepwise, or ratcheted, curves (such as the one in Figure D.2). But the stepwise representation is what gives the basic intuition for interpreting consumer surplus as overall economic benefit to buyers in any market.

Figure D.3
Consumer Surplus with Smooth Demand

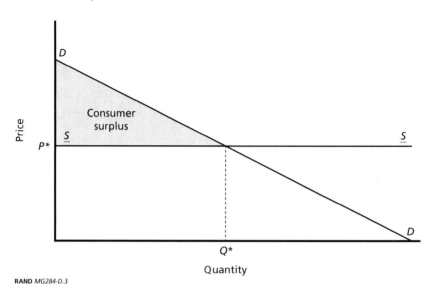

Applying the Economic Analysis Framework

What are the prospects for the overall size of the GPS-related market as a result of Galileo's appearance? These are the issues:

1. What underlying cost characteristics of Galileo will contribute to its being voluntarily adopted by GNSS users?
2. If Galileo would be voluntarily adopted at a zero surcharge to Galileo users (as in U.S. policy toward GPS), how high a surcharge can Galileo management impose and still have Galileo be adopted voluntarily?
3. If Galileo management imposes regulations on Galileo's use, what is the economic impact of such regulations for industry and users (the market for GPS services and receivers)?

We first consider a case in which Galileo is solely a supplement to GPS—that is, U.S. GNSS users choose either a GPS-only system or a combined GPS-Galileo system. We then consider the more complex case, in which all three options are available: GPS-only, Galileo-only, and the combined GPS-Galileo.

For ease of exposition, we have assumed that there is a single U.S. market for GNSS (i.e., only one market segment) in a single representative time period. The analysis can thus be illustrated in a single demand diagram (Figure D.4) that applies to a representative U.S. consumer of GNSS in a GPS-only world.

We normalize the quantity (Q) at unity, as well as the price (P) that users must pay to receive the services. Since there is no U.S surcharge for using GPS services, this price represents the market price of the equipment and/or service. The supply curve is flat, in accordance with the fact that this is being analyzed from the standpoint of a single representative U.S. consumer who has no control over the

Figure D.4
Diagram of GPS-Only Market

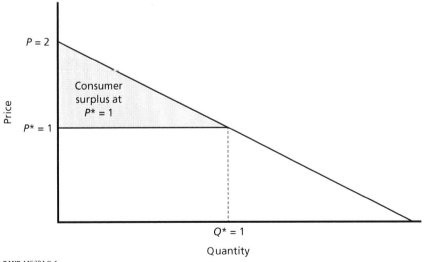

price of GPS services. Here, we assume a price elasticity of demand[4] of −1 at the market-equilibrium price and quantity, and a linear demand curve. Given these assumptions, the demand curve is

$$P = 2 - Q \qquad (1)$$

and the numerical value of consumer surplus, calculated from the triangular area shown in Figure D.4, is 0.5. We represent the introduction of a Galileo system into this market as an outward and upward shift in the U.S demand curve, which is shown in Figure D.5. As described above, the demand curve represents the value of each increment of GNSS use to GNSS users. Thus, in economic terms, the

Figure D.5
Diagram of GPS and Galileo Market

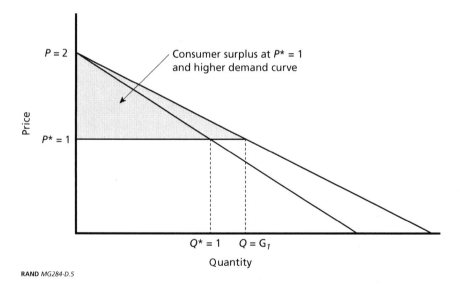

---

[4] "The measure of the responsiveness of supply and demand to changes in price is their elasticity. Elasticity is calculated as the ratio of the percentage change in the quantity demanded or supplied to the percentage change in price. Thus, if the price of a commodity decreases by 10 percent, and the sales of it consequently increase by 20 percent, the elasticity of demand for that commodity is said to be 2" (*Britannica Online*).

shift in the demand curve represents an increase of the user-perceived value of a combined GPS-Galileo system over a GPS-only system.

The height of the demand curve is the U.S. users' *willingness-to-pay* for each increment of GNSS. We suggest that one way the value of GNSS use can increase (i.e., the demand curve can shift upward) with the addition of a Galileo system is if users can avoid the cost of another augmentation system by switching from a GPS-only system to a combined GPS-Galileo system. Their *willingness-to-pay* for the combined system will be higher than their *willingness-to-pay* for the single system, simply because they can avoid other costs with the combined system.

At this point in the analysis, we need to first describe two possible alternatives facing U.S. consumers. In one (Figure D.4), U.S. consumers are purchasing GNSS in a GPS-only world. In the other (Figure D.5), U.S. consumers are purchasing GNSS in a world that includes both GPS and Galileo. The question being addressed is how high can the price paid by U.S. consumers rise in the combined GPS-Galileo situation before U.S. consumers prefer only GPS. If they prefer only GPS, there would be an incentive to keep the markets segmented.

We represent this change in the demand curve as g, the percentage growth in demand for GNSS that results from the addition of the Galileo system, a multiplicative shift of $G = (1 + g)$ in each quantity. In other words, it is a horizontal multiplication of the demand curve by a factor $G = (1 + g)$. A shift of zero—the demand curve is unchanged—means there is no economic benefit (no change in the consumer surplus) as a result of the addition of Galileo.

The point $P=1, Q=G$ is on this new demand curve, and the elasticity of demand at this point remains at -1. For any given price P of GNSS services, consumer surplus is

$$\frac{G(2-P)^2}{2}.$$

(2)

(Note that, for $P=1, Q=1$, consumer surplus equals 0.5, consistent with our earlier observation on consumer surplus in the GPS-only case.)

We now ask: How high can the price go before U.S. GNSS users prefer *not* to use the combined system—that is, before they would be better off paying a price of unity and getting only the benefits associated with a GPS-only system? This maximum is expressed as

$$2 - \frac{1}{\sqrt{G}} \approx 1 + \frac{g}{2}. \tag{3}$$

Given our assumptions, we find that the price can grow by about half the rate at which demand grew without users losing their preference for the combined system. At a higher price, however, they will prefer to stick with GPS-only. (This result is *critically* dependent on the price elasticity assumption.) This price increase includes both whatever real cost increase in receivers is required to receive the combined GPS-Galileo signal *and* whatever surcharge Galileo management may impose.

Therefore, if the proportionate cost increase of combined GPS-Galileo receivers is more than half the increase in the demand caused by the addition of Galileo, U.S. consumers will prefer to remain with GPS only. (Again, this result is critically dependent on the price elasticity assumption. This warning applies to all further results and will not be repeated.)

So, one critical market test for Galileo is this relation of the cost of combined receivers/service to the benefits of improved technical or operational performance. Let the cost of a combined GPS-Galileo receiver be c ($c > 1$, since 1 is the cost of a GPS-only receiver in this analysis). As just noted, if

$$c > 2 - \frac{1}{\sqrt{G}} \tag{4}$$

U.S. consumers will not voluntarily adopt Galileo.

Now, let us say that Galileo management imposes a surcharge s on each combined receiver/service. We might ask how total surcharge revenues vary with the surcharge. The price is now $(s + c)$, and surcharge revenues are

$$sQ = s\left(2 - \left(c + s\right)\right)G. \tag{5}$$

If GNSS users were *compelled* to use the combined system, the revenue-maximizing s would be $(2 - c)/2$. However, users will prefer to switch to a GPS-only system if the receiver price is more than $[2 - 1/(G)^{1/2}]$ (i.e., if the surcharge is more than $[2 - 1/(G)^{1/2} - c]$). This option to switch to a GPS-only system is, in fact, likely to constrain surcharge revenues. Say that the switch-to-GPS-only option does indeed constrain the surcharge revenues that are possible, in the sense that

$$2 - \frac{1}{\sqrt{G}} - c < \frac{\left(2 - c\right)}{2}. \tag{6}$$

Galileo managers still may have an option to increase revenues beyond the level achieved at a surcharge of $[2 - 1/(G)^{1/2} - c]$.

One can imagine various scenarios in which both GPS and Galileo are sold to U.S. consumers. For example, there may be a fraction of the market—say, e—for which Galileo use is mandated. Galileo's maximum surcharge revenues from only this part of the market would still be achieved at a surcharge level of $(2 - c)/2$, although it would only be e times as high. For reasonable parameter values, this strategy may lead to higher revenues from this limited part of the market (those that were compelled to use) than could be gained from the total market, with surcharges constrained by the switching option. Since this policy would lower the use of GPS signals in the e fraction of the market as a result of its higher overall price, it would reduce sales of GPS receivers from what they would be in an unconstrained market—a case in which requirements imposed by Galileo management to increase surcharge revenues reduce the market for GPS receivers—and could be considered predatory. U.S. policymakers will want to analyze actual market prospects to determine whether such predation is likely and to include these considerations in their negotiations.

This is only one way to represent various Galileo management policies. It captures the essence of a concern about predation/destructive competition: that the GPS receiver industry is reduced through arbitrary regulation rather than through constructive (fair-and-square) competition. The other ways to represent such policies should be explored as well.

Bibliography

Airways New Zealand, "Global Navigation Satellite Systems: Receiver Autonomous Integrity Monitoring (RAIM) Prediction," n.d., online at http://www.airways.co.nz/airways_Services/global_nav_2.asp.

Beichman, Arnold, "Space Wars," *The Washington Times*, April 8, 2002, p. 17.

Benedicto, J., S. E. Dinwiddy, G. Gatti, R. Lucas, and M. Lugert, *Galileo: Satellite System Design and Technology Developments*, Noordwijk, The Netherlands: European Space Agency, November 2000.

Berger, Samuel, Jacob Lew, and Neal Lane, Memorandum for the Secretary of State, the Secretary of Defense, and the Secretary of Transportation, "Implementation of GPS Second and Third Civil Signal," The White House, Washington, DC, January 15, 1999.

Betz, John (MITRE), interview with author, April 2, 2003.

Braibanti, Ralph L. (U.S. Department of State), "The Global Positioning System: International Cooperation," briefing charts presented during ION GPS 2002 plenary session, September 6, 2002.

Brown, Alison (President and Chief Executive Officer, NAVSYS), interview with author, February 12, 2003.

Bruns, Markus, "Status of the Galileo Program," issue paper presented at ICAO GNSS panel meeting, San Antonio, TX, ICAO GNSSP WHL IP-4, October 16–25, 2002.

Ciganer, Ann (Vice President, Trimble), interviews with author, February 13, 2003, and April 8, 2003.

Clark, John E., "GPS Modernization Update," briefing charts presented at ION 2002 National Technical Meeting: Integrating Technology, San Diego, CA, January 28–30, 2002.

Divis, Dee Ann, "GPS, Galileo Draw Closer," *GPS World*, November 2002.

Enge, Per (Stanford University), "A Global Challenge—Protect the GNSS Noise Floor," presented at ION GPS 2002 plenary session, Portland, OR, September 2002.

European Commission, *GALILEO Finally Takes Off*, Brussels, Belgium, ip/02/478, March 26, 2002, online at http://www.heise.de/english/newsticker/news/26085.

European Commission, *Galileo: The European Programme for Global Navigation Services*, Noordwijk, The Netherlands: European Space Agency, BR-186, May 2002.

Fontana, Richard D., Wai Cheung, and Tom Stansell, "The Modernized L2 Civil Signal: Leaping Forward in the 21st Century," *GPS World*, September 2001.

Frost & Sullivan, "The Growing Importance of Electronic Navigation Equipment," executive summary in Frost & Sullivan, *European Electronic Navigation Equipment Markets*, March 31, 1998.

Frost & Sullivan, "Introduction to the North American GPS Market," executive summary in Frost & Sullivan, *The North American GPS Market*, 2000.

Fry, George (Aviso Micro Technology), *GPS '99: A Commercial Market Analysis*, Forward Concepts, 1998.

Fyfe, Peter, et al., "GPS and Galileo—Interoperability for Civil Aviation Applications," *ION GPS 2002*, 2002, pp. 289–302.

Galileo Services and Architecture, Attachment 1 to Markus Bruns, "Status of the Galileo Program," issue paper presented at ICAO GNSS panel meeting, San Antonio, TX, ICAO GNSSP WHL IP-4, October 16–25, 2002.

Gholz, Eugene, *National Security Space Policy in the U.S. and Europe: Trends and Choices*, conference report, University of Kentucky, MIT Security Studies Program, October 2002.

"GLONASS, GPS and Galileo: A Multi-Expert Interview," in *ION Newsletter*, Vol. 13, No. 1, Spring 2003.

GPS ORD'99 (Operational Requirements Document [ORD] AFSPC/ACC 003-92-I/II/III for Global Positioning System [GPS]), El Segundo, CA: AFSC/ACC, 1999.

GPS Partnership Council, "Planning the Next Generation of GPS," briefing charts presented at meeting hosted by Armed Forces Communications and Electronics Association, Los Angeles Chapter (AFCEA-LA), El Segundo, CA, September 6, 2001.

"GPS Takes on Galileo," *Aerospace America*, August 2003.

GPS World Buyers Guide, June 2002.

GPS World Receiver Survey, January 2003.

Grejner-Brzezinska, Dorota (Ohio State University), "Telegeoinformatics: Positioning and Tracking Approaches and Technologies," n.d.

Hasik, James, and Michael Russel Rip, "An Evaluation of the Military Benefits of the Galileo System," in *ION GPS 2002*, 2002, pp. 320–329.

Hatch, Ron, Jaewoo Jung, Per Enge, and B. Pervan, "Civilian GPS: The Benefits of Three Frequencies," n.d., online at www.navcomtech.com/docs/CivilianGPS.pdf.

Hein, Guenter W. (Member of EC Signal Task Force), *GPS-Galileo Compatibility & Interoperability*, briefing charts presented at ION GPS 2002 plenary session, Portland, OR, September 2002.

Hein, Guenter W., et al., "Status of Galileo Frequency and Signal Design," Members of the Galileo Signal Task Force of the European Commission, Brussels, in *ION GPS 2002*, 2002, pp. 266–277.

Heinrichs, Guenter, "Receiver Architecture Synergies Between Future GPS/Galileo and UMTS/IMT-2000," *IEEE*, 0-7803-7467, March 2002, pp. 1602–1606.

Holmes, J. K., and S. Raghavan (The Aerospace Corporation), "GPS Signal Modernization Updated Summary," in *ION 58th Annual Meeting/CIGTF 21st Guidance Test Symposium*, 2002, pp. 544–545.

International Telecommunication Union, Radiocommunication Study Groups, United States of America, Information Paper, "Civil, Commercial and Consumer Use of the Global Positioning System (GPS)," Document 8D/58-E, April 16, 1998.

Issler, Jean-Luc, *Galileo, GPS/Galileo Interference Computation Models*, Centre National D'Etudes Spatiales (Departement Radio Navigation), DTS/AE/TTL/RN-2002-076, October 17, 2002.

Jeans, Bob, *Benefits and Issues Related to the Combination of Galileo and GPS*, working paper presented by Markus Bruns at 11th Air Navigation Conference, San Antonio, TX, ICAO GNSSP WHL WP-8, October 16, 2002.

Jones, Ray, et al., "GALILEO Augmentations for Precise Positioning Applications," in *ION GPS 2002*, 2002, pp. 1704–1715.

Kayton, Myron, review of Pratap Misra and Per Enge, *Global Positioning System: Signals, Measurements and Performance*, online at http://www.navtechgps.com/supply/2500.asp.

Kim, Jason Y., *The Global Positioning System: A Worldwide Information Utility*, Executive Secretariat, Interagency GPS Executive Board (IGEB), November 12, 2002.

Kim, Jason Y., and Ralph Braibanti, *GPS-Galileo Negotiations: Commercial Issues at Stake*, Office of Space Commercialization, U.S. Department of Commerce, Office of Space & Advanced Technology, U.S. Department of State, March 21, 2002.

Kovach, Karl L., "GNSS Signal Compatibility and Interoperability, GPS Person's View," briefing charts presented at ION GPS 2002, Portland, OR, September 24–27, 2002.

Kovach, Karl L., and Karen L. Van Dyke, "GPS in 10 Years," *Microwave Journal*, Vol. 41, No. 2, February 1998, pp. 22–29.

Lachapelle, Gerard (ION Western Region Vice President and head of the Geomatics Engineering Department at the University of Calgary, Alberta, Canada), interview with author, 2003.

Lancop, Robert (Civil Matters Branch, Competition Bureau), "Competitive Safeguards: The Role of the Competition Bureau in the Evolving Telecom Market," notes for an address, Insight Conferences and Report on Business Conference, Regulating International Telecommunications, Toronto, Ontario, Canada, November 5, 1997, online at http://cb-bc.gc.ca/epic/internet/incb-bc.nsf/vwGerneratedInterE/ct01439e.html.

Lavrakas, John W. (Overlook Systems Technologies, Inc.), "Planning the Future of GPS," in *ION 58th Annual Meeting/CIGTF 21st Guidance Test Symposium*, 2002, pp. 538–544.

Lee, Jennifer, "Europe Plans to Compete with U.S. Satellite Network," NYTimes.com, November 26, 2001.

Leick, Alfred, *GPS—A National Asset and Treasure*, Department of Spatial Information, University of Maine, 1999.

Loverro, Douglas (Colonel, USAF), interview with author, October 21, 2002.

Maine, Kristine P., and Thomas A. Stansell, Jr., "Aviation Use of RNSS in the 1215–1300 MHz Band," *ION 58th Annual Meeting/CIGTF 21st Guidance Test Symposium*, 2002, pp. 569–580.

Martin-Neira, M., P. Colmenarejo, G. Ruffini, and C. Serra, "Altimetry Precision of 1 cm over a Pond Using the Wide-Lane Carrier Phase of GPS Reflected Signals," *Canadian Journal of Remote Sensing*, Vol. 28, No. 3, 2002, pp. 394–403.

Mastracci, Claudio (Director of Applications, ESA), "The GALILEO System and Its Status," briefing charts presented at Galileo Industry Day, March 18, 2002, online at http://europa.eu.int/comm/dgs/energy_ transport/galileo/partners/private_infoday_en.htm#presentations.

McCall, Gene (Chief Scientist/U.S. Air Force Space Command), interview with author, December 16, 2002.

National Imagery and Mapping Agency (NIMA) Satellite Geodesy Team, "Current GPS Satellite Data," n.d., online at http://www.nima.mil/ GandG/sathtml/satinfo.html.

Navigate Consortium, *Navigate Consortium Position Paper*, July 8, 2002, online at http://www.navigateconsortium.it/page_14.html.

Ochieng, W. Y., et al., "Potential Performance Levels of a Combined Galileo/GPS Navigation System," *The Journal of Navigation*, Vol. 54, No. 2, n.d., pp. 185–197.

O'Donnell, Matt, et al., "A Study of Galileo Performance—GPS Interoperability and Discriminators for Urban and Indoor Environments," in *ION GPS 2002*, 2002, pp. 2160–2172.

O'Keefe, K., S. Ryan, and G. Lachapelle, "Global Availability and Reliability Assessment of the GPS and Galileo Global Navigation Satellite Systems," *Canadian Aeronautics and Space Journal*, Vol. 48, No. 2, June 2002, pp. 123–132.

Onidi, Oliver (Head of Galileo Unit, European Commission), "Galileo and GPS: True Added Value Through Interoperability," briefing charts presented at ION GPS 2002, Portland, OR, 2002.

Pace, Scott, et al., *The Global Positioning System—Assessing National Policies*, Santa Monica, CA: RAND Corporation, MR-614-OSTP, 1995.

Poulter, Tony (PricewaterhouseCoopers), *Galileo—The Commercial Structure and Revenue Opportunity*, briefing charts presented at Galileo Industry Day, March 18, 2002, online at http://europa.eu.int/comm/dgs/energy_transport/galileo/partners/private_infoday_en.htm.

Radovanovic, Robert S., and Naser El-Sheimy (Department of Geomatics Engineering, University of Calgary), "Using Optimal GNSS Multi-Frequency Carrier Phase Combinations for Precise Kinematic Positioning," *ION 58th Annual Meeting/CIGTF 21st Guidance Test Symposium*, 2002, pp. 223–230.

Reaser, Rick (Colonel, USAF), conversations and e-mail correspondence with author between October 2002 and February 2003.

Reaser, Rick, "Navstar Global Positioning System—GALILEO," briefing charts, Navstar GPS Joint Program Office, January 6 and 10, 2003.

Roturier, Benoit, "The Concept of Using Combinations of Independent Constellations (e.g., GPS, GLONASS and Galileo)," draft V3.2 of working paper, ICAO GNSS panel meeting, San Antonio, TX, ICAO GNSSP WGW Flimsy 10, October 22, 2002.

Roturier, Benoit, & Drafting Group No. 8, "The Concept of Using Combinations of Independent Constellations (e.g., GPS, GLONASS and Galileo)," draft V2 of working paper, presented by Michel Calvet at ICAO GNSS panel meeting, San Antonio, TX, GNSSP WHL WP3, October 16, 2002.

Sabathier, Vincent, and Harvey Sapolsky, *National Security Space Policy in the U.S. and Europe: Trends and Choices*, final report, Cambridge, MA: Massachusetts Institute of Technology, October 2002.

Stansell, Tom (President, Stansell Consulting), interview with author, March 4, 2003.

Turner, Dave, et al., *GPS and Galileo—Compatibility or Interoperability? A Hierarchical Assessment of Time, Geodesy, and Signal Structure Options for Civil GNSS Services*, The Aerospace Corporation, June 2002.

U.S. Department of Commerce, International Trade Administration, Office of Telecommunications, *Global Positioning System: Market Projections and Trends in the Newest Global Information Utility*, September 1998.

U.S. Department of Commerce, Office of Space Commercialization, *Promoting Commercial Interests in GPS*, July 2002.

U.S. Department of State, Office of the Spokesman, "U.S. Global Positioning System and European Galileo System," media note, Washington, DC, March 7, 2002, online at http://www.state.gov/r/pa/prs/ps/2002/8673.htm.

Verhagen, Sandra (Delft University of Technology, Department of Mathematical Geodesy and Positioning), "Performance Analysis of GPS, Galileo and Integrated GPS-Galileo," in *ION GPS 2002*, 2002, pp. 2208–2215.

Wall, Robert, and Craig Covault, "Eroding GPS Worries Pentagon," *Aviation Week & Space Technology*, November 4, 2002.

Wilson, Andrew (ed.), "Galileo—The European Programme for Global Navigation Services," briefing charts, Noordwijk, The Netherlands: ESA, BR-186, ISBN 92-9092-730-5, May 2002.

Wolfowitz, Paul (U.S. Deputy Secretary of Defense), letter to Geoffrey Hoon, Ministry of Defense, United Kingdom, dated December 1, 2001.